Table of Contents

1. Introduction 1
   - 1.1 The pioneers of the future 4
   - 1.2 The laws of the cosmos 10
   - 1.3 The demise of the ether 14
2. Einstein and the theory of relativity 19
   - 2.1 The special theory of relativity 20
     - 2.1.1 The Einstein Postulates 24
     - 2.1.2 Red Wine, Trains and the Pizza Bet 30
     - 2.1.3 Does a second always last a second? 35
     - 2.1.4 Does one metre always measure one metre? 39
     - 2.1.5 The garage paradox 43
     - 2.1.6 Of Zombies, the conductor and simultaneity 47
     - 2.1.7 Meteorites, astronauts and mass 50
     - 2.1.8 Einstein's formula "$E = mc^2$" 52
   - 2.2 The General Theory of Relativity 57
     - 2.2.1 The curved space-time in four dimensions 58
     - 2.2.2 The passage of time and gravity 67
     - 2.2.3 The Shadows of Gravity 72
     - 2.2.4 The limits of the universe 80
     - 2.2.5 Escape from the universe 89
     - 2.2.6 The Einstein-Rosen Bridge 95
     - 2.2.7 Time travel and the fourth dimension 101
     - 2.2.8 Tachyons and the past 114
     - 2.2.9 Gödel's formula: time at the end of the universe 120
     - 2.2.10 The black hole 123
     - 2.2.11 Antigravity and dark energy 136
3. The quantum nightmare 143
   - 3.1 The quantising of light, space and time 144
   - 3.2 The Eerie World of Quantum 152
   - 3.3 Einstein, Schrödinger and the half-dead cat 162
   - 3.4 The Copenhagen Interpretation 172
   - 3.5 The quantum tunnelling effect 176
   - 3.6 The eerie long-distance effect 182
   - 3.7 The quantum nightmare in a vacuum 187
   - 3.8 Antimatter 192

- 3.9 The Age of Antimatter 204
- 3.10 The absolute zero point 208
- 3.11 The Higgs particle 218

4. The theory of everything 224

- 4.1 The search for the blueprint 224
- 4.2 The String Theory 230
- 4.3 Strings - The smallest building blocks of matter 234

5. Outlook for the 3rd millennium 239

- 5.1 The 22nd century 239
- 5.2 What if 243
- 5.3 What came before the big bang? 245
- 5.4 All coincidence? 246
- 5.5 The Century Puzzles of Physics 250
  - 5.5.1 The 18 unknowns 251
  - 5.5.2 The Mystery of Gravity 251
  - 5.5.3 Why a universe of matter? 252
  - 5.5.4 Where does mass come from? 252
  - 5.5.5 How many dimensions are there? 253
  - 5.5.6 Are there other universes? 254
  - 5.5.7 Does dark energy exist? 254
  - 5.5.8 What happens in the black hole? 255
  - 5.5.9 Where does life come from? 255
  - 5.5.10 Mysteries of Quantum Physics 256
  - 5.5.11 Is mathematics discovered or invented? 256
  - 5.5.12 What next? 257

1. Introduction

There is a secret world of science very few people know about. A science which has discovered how nature ticks and has explored the very fabric of its innermost core. A science which is revolutionising our view of the world but which is as real as you and I. I am talking about modern physics, the science which was discovered by ingenious researchers such as Albert Einstein over a hundred years ago and which has changed our world view for ever. Yet today there is hardly anyone who has not actually studied physics who can possibly guess how astonishing and extraordinary nature is beyond our intuition. Or would you have known that Newton's formulae, which are today taught in schools, are in fact wrong? Or that time on Earth passes more slowly than on the Moon? Or that our universe consists of at least four dimensions? Or that time travel is not a pipe dream?

The key to spectacular technologies that surpass the creative spirit of the best science fiction authors can be found in modern physics. These are technologies which are researched and developed at renown universities and research centres. The whole gamut ranges from antimatter through quantum computers to futuristic-looking time machines and parallel universes.

Time machines? Parallel universes?

Yes, you read right! When you think of it, the author must be at least as crazy as the publishing company which publishes such things, and you will probably feel like I did - fifteen years ago. But I can assure you of two things:

The last hundred years of research have turned our world view on its head. Nothing is as it was before Einstein and company. Apart from the fact that even in the year 2018 only a very few people know how extraordinary our world actually is, on no matter what scale.

Secondly - and this is by far the most astonishing fact you will find on the second page of a book about modern physics - all these technologies and phenomena are as real as you and I. I think, therefore I am. The best researchers and scientists are working on converting the fundamentals newly acquired from the theory of relativity and quantum physics to practical applications. Or they are concerned to explain why everything that moves faster than the speed of light disappears into the past. Or why the outcome of an experiment in the microcosmos depends on whether someone is looking at it or not. I could give you twenty examples of this right now and in doing so I could already turn most of your world view upside down. However, the theory of relativity and quantum physics are only the first steps towards an entirely new concept of how the universe works and ticks. Both these revolutionary theories lead on to the theory of everything, a unique theory from which all natural laws can be derived.

These are not primarily new technologies but rather represent whole new understanding of the universe, nature and the world. You will certainly agree with me when I say that each year everyone becomes one year older. If not mentally, then at least physically. This event tends to be celebrated with much merriment, with cakes, presents and a good measure of jollity. You will also readily agree that it does not matter whether a person spends the year in the USA, Europe or in a space station on the moon. Finally, time passes everywhere and just as quickly for everyone – even if work days are felt to be at least twice as long as the weekend or holidays.

The Strange Universe: Einstein, Quantum Physics and the ToE

Now when I also tell you that I do not leave my house through the wall but through doors or windows, you will nod in agreement – or you will wonder whether an attempt is being made here to fill the pages with trivial words of wisdom.

Until Albert Einstein shook the foundations of physics with his theory of relativity, no sensible person would have even considered questioning seriously any of his assertions. In fact you have to sympathise with the idea that it would be extremely naive to agree with these three assertions. None of them is true. No matter how we look at it. Our perception of reality is like looking through the rose-tinted glasses of a newly in love couple. Instead let us put on our reading glasses and venture into the first chapter of the world plan, the primordial script of things, right up until the fantastic, mysterious and rare discoveries which this world holds for us.

It is time to open our eyes.

The world is far more complex than the science of our great grandfathers could ever have imagined. When you think that there cannot be much out there that is not known to us, with the exception of sensation-seeking authors, you are no further forward than the position of science at the end of the 19th century. Even then massive mistakes were being made. The world which we perceive in our everyday lives is only an appearance, a simplification of reality. It is as if our senses were covered with a veil. Yet the world behind this veil is immense and unimaginably different. Just consider the starlit sky on a clear summer night, filled with an unbelievable number of tiny, brightly shining dots which mark the firmament. The highlight of any romantic evening. And a journey in time into ancient epochs. The light from the stars, which shine incomprehensible distances, has travelled over millions and billions of years to reach the Earth. All these constellations and formations originate from the distant past. No-one knows what is happening right at this minute far out in the universe. No astronomer. No telescope. No space probe. The starry sky is only an illusion or reality. A glimpse into the past, a witness to a world that may have existed in a long forgotten world. No-one knows whether the stars we see today actually still exist.

And this is only the beginning of a breathtaking journey into the origin of this world. Into a world which so far no-one knows hardly anything about. Into a world which no-one understands.

Let us turn back the clock, for everything began over three hundred years ago in England.

1.1 The pioneers of the future

The 5th of July 1687 was to go down in history as the date on which Isaac Newton changed the world. It was the date on which he released the falling apple from the tyranny of capricious arbitrariness and subjected it and all matter to the laws of nature. It was the date on which he published his "Principa Mathematica". A work in three parts in which he formulated, among other things, the mathematical laws of movement and gravity[1]. With them he destroyed the prevailing school of thought according to which the laws of nature on Earth and in the sky are different by combining the research of Galileo Galilei on acceleration and Kepler's laws of planetary movement into a theory of gravity. Newton formulated

---

[1] Gravity = force of attraction = force of gravity (the terms are used synonymously and, to put it simply, designate the force which keeps us on the Earth)

laws of movement and gravity which form the basis of classical mechanics. He discovered that gravity was the cause of planetary movements and made major contributions to the science of optics. He was, moreover, a brilliant mathematician and co-inventor of calculus. Newton is considered one of the most influential and significant scientists of all time. His research gave us an insight into universal laws of nature and hence to the knowledge that all things in the universe obey the same principles.

Newton devoted a large part of his life to the question of the essence of time and space and our origins. Newton understood the universe to be a gigantic clock bringing order to an otherwise chaotic state through time. He considered time to be the guardian of things responsible for ensuring that all events take their predetermined course. The house cannot burn until lightning has struck the roof.

The English philosopher and physicist considered fate to be predetermined. He was convinced that the forward march of time is unstoppable and that the past, present and future lie in a continuous sequence and are therefore unchangeable. He stamped an absolute world view on contemporary thinking, a world view which resists free will and sees humanity as helpless floss drifting in the flow of time. Newton's theories manifested the slits of wisdom on the declaration of unfree will. Over 300 years ago Newton represented a world view which most of our fellow beings still hold. Yet troubled times were afoot.

When the first factory chimneys began to pollute the air with soot and smoke, the pioneers of a fundamental revolution came onto the world scene. They were born in the ensuing industrialisation. Yet their legacy would one day become much more far-reaching.

They had come to overturn age-old wisdoms and truths defended for generations. They fought with matters of fact that had gathered dust. They destroyed the world view that is predictable in nature, the clock which follows a primordial cycle in absolute uniformity. They changed the world with their life's work. For these pioneers had discovered something. A concept of time and space. A new reality. New dimensions which relativised Newton's legacy. A closely guarded secret was buried in the old science textbooks. A secret whose veil was not lifted until Einstein, Planck, Maxwell and company were called into action.

What no-one knew for over 200 years: the formulae of Newton mechanics, which are today taught in schools all over the world, are fundamentally wrong. These are the formulae with which satellites have been put into a stable orbit around the Earth, men have been sent to the moon or unmanned space probes have been sent beyond the solar system. It is also these formulae with which the police work out your speed, parachutists plan the time they open their parachutes, or the trajectory of a long-range ballistic missile is calculated.

So is half the technical world based on formulae which are wrong? And if this is the case, why is this hardly given a mention in schools and educational courses? And why, then, don't satellites fall out of the sky?

The first question can be answered with two letters.

The second question therefore derives from the fact that although this falsehood had been known almost a century ago, very few people had even heard about it. Even teachers are people.

There may be a number of reasons for this. The complexity of the subject is certainly a contributory factor. Proving that Newton's formulae are actually wrong is a very difficult thing to do. Even more difficult than Newton's formulae already are themselves. What is more decisive, however, is the fact that the formulae of Newton mechanics produce very good results in everyday life and are not therefore totally wrong. Physicists tend in this connection to talk about an approximation of reality. If you find yourself at a police checkpoint and you explain to the friendly police officers that you have only drunk one glass of alcohol, by which you mean a good bottle of red wine, the one lying next to the three empty bottles, that is not totally wrong either. The legendary fall of the apple can be described by Newton's formulae just as accurately as the speed of a vehicle. If you drive your car too quickly on a national highway, the accuracy is also sufficient to calculate your speed very precisely. But the result never corresponds exactly to reality. However, the deviation is very small in this speed range and is therefore hardly measurable in any case. For this reason the satellite also remains within its orbit and the probes reach planets and moons reliably. Nevertheless, the error can in no way be ignored if you want to prevent technological progress from ending at this point. For the effects of the discrepancy between the Newton formulae and reality can be seen surprisingly quickly in other applications that had become commonplace.

If Albert Einstein had not discovered the theory of relativity (and no-one else), you could not blame the taxi driver if he got lost in the maze of streets of the city. For even in GPS navigation systems Newton mechanics are no longer sufficient to be guided safely and accurately through the streets (and not go into a post or street light).

GPS satellites determine the position in which you calculate your address from phased signals. In the GPS satellite and in the GPS unit in your car there are high-precision clocks which are synchronised, i.e. they are coordinated with each other. Therefore they should actually show the same time. The satellites orbit the earth at an altitude of approximately 23,000 kilometres, where time strangely passes faster than on earth. An effect which has been discovered, explained and has since been confirmed in numerous experiments. Thus if no-one had discovered the theory of relativity until the discovery of GPS navigation, the GPS would miss its position by around one kilometre after travelling for approximately one hour. It is therefore conceivable that the friendly female voice will explain to you that you have reached the football ground but you have parked the car somewhere in the ditch.

Newton's formulae are no doubt brilliant and are an epoch-making achievement. In the 17th century they marked a huge leap in scientific progress and formed the basis of classic physics. For the first time nature could be described quite extensively in mathematical terms. For example, the law of gravity could be applied just as practically to falling apples as to calculating the orbits of the planets and the stars. Even in the 21st century many technical applications and constructions are based on these mechanics. Yet these formulae are only what many other formulae of our time also appear to be[2]. An approximation of reality. A simplification of the reality which is not found in everyday life, and perhaps not even suspected. However, this difference between reality and the formula suddenly becomes significant with the technological progress over decades and centuries. In Newton's time no-one would have been able to conduct an experiment that would have proved that the formulae were inaccurate. However, it

---

[2] ... and in certain circumstances cannot even guess at.

is a fact that not even the accuracy of GPS satellites is sufficient to calculate position data any longer. Newton's formulae are by no means the only rules of physics that were intended to evolve into an unimaginably more comprehensive theory. The beginning of the twentieth century rather marked the starting point for lasting changes that were to surpass everything that humanity had learned and discovered in the last five hundred years about the condition of the world. It was to become an established fact that other dimensions have to be negotiated in order to explain nature, dimensions which are difficult to access in our daily lives. If we are ever to really understand our world we must penetrate realms that are alien to us. Alien in terms of their existence. Alien in terms of their essence. Alien in terms of their principles.

In the last 150 years much has changed in many respects. The tight constraints of the invisible illusion were released. The knowledge that had been cemented over whole generations was shaken. The tide had turned. The flow had lost its power. A new science suddenly opened doors and paths to a reality that had come to be expected across every previous epoch. The mechanism for enormous change had been set in train. The very essence of humanity, its existence, possibilities and mission were to see decisive changes, giving our world view new dimensions.

The end of the 19th century heralded in the demise of classical physics. The quantum and relativity theories were thrown into the spotlight and replaced the rigid views of Newton. They discovered the essence of time and unleashed free will from the primordial dictate. They paved the way towards an incredible future with technologies and knowledge which would still have been considered absolutely unthinkable at the time the railway came into existence.

Immerse yourself in the fascinating secrets of our world. Popular terms such as antimatter, space time, time travel, string theory, theory of relativity or quantum mechanics take on a whole new, tangible and realistic meaning in this book. For modern physics brings us not only an abundance of new technologies but a whole new understanding of the world, a new world view. Merely the fact that the result of an experiment depends on whether someone is observing it or not, or that the microcosmos is based on probabilities, provides sufficient ammunition for permanently shattering our world view.

1.2 The laws of the cosmos

The water in the sea, the flowers in the garden, the petrol at the filling station, and of course also the petrol pump – all these things and all matter known to us consists of tiny atoms, so-called atoms. A single dust particle consists of billions of atoms. Each atom consists of even smaller particles, namely the electrons, neutrons and protons, the so-called elementary particles. For example, these elementary particles can be represented as tiny balls. Electrons are negatively charged elementary particles which circle around the nucleus of the atom in a kind of bowl in a simple representation, and therefore form its envelope. This is rather similar to the way in which the earth and the other planets of the solar system encircle the sun. The core of the atom, the atomic nucleus, consists of the protons, positively charged particles, and neutrons, which are uncharged (neutral). We describe electrons, protons and neutrons as elementary particles. In fact these particles do not form the fundamental component of matter, for there are even smaller elements, the quarks. Each of our elementary particles therefore consist in turn of even smaller particles[3]. Quarks are in many respects rather rare particles, for although it could be proved that the elementary particles consist of different quarks, it seems as if quarks would not exist outside elementary particles. Taking the analogy of the solar system, this means, for example, that the earth consists of boulders and rocks, but these do not exist outside our planet except in or on another planet. At least no isolated quarks have so far been observed. And a good thing too. Some scientists fear that the so-called "strange-quark" is extremely dangerous outside any bond. So it could absorb matter and convert to rare matter, whilst other matter apparently dissolves into nothing. A kind of black hole in particulate form.

You may perhaps wonder whether the quarks are now the smallest constituents of our world, or whether they in turn consist of even smaller elements?

At the end of the 1960's the Americans changed the world by putting the first human being on another celestial body with the moon landing. A remorseless war was raging in Vietnam. In Woodstock the hippies were enjoying themselves at the height of the era of sex, drugs & rock'n'roll. At that time the Italian physicist Gabriel Veneziano discovered the smallest components of the world from one formula, the strings. This was a discovery which was to form the basis of the theory of everything. As early as the beginning of the twentieth century researchers had begun to build a bridge between the mysteries of physics and to combine all the problems and questions of physics under the roof of one single

---

[3] Strictly speaking this only applies to the protons and neutrons, since electrons belong to the so-called "leptons" which, together with quarks and force transmission particles (gauge bosons) form the building blocks of matter.

theory. In one of these theories – the theory of everything – there are the quarks and generally all the particles deriving from even smaller elements, the strings. The mathematics of this string theory is highly complicated and the theory is in and of itself so fundamental that to date we have been unable to determine experimentally whether it is correct. The theory of everything has set itself the goal which Einstein was unable to achieve throughout his life: To combine the quantum theory and the theory of relativity under the roof of one single theory. What sounded like academic banter turned out to be the physics problem of the century. Edward Witten even described the string theory as the research field of the 22nd century, a field which by pure accident has manifested itself in our present time .

Quantum physics is one of the cornerstones of modern physics and deals with the microcosmos and elementary particles. As a rule of thumb this includes everything which is so small that it cannot be seen with the most powerful microscopes and therefore lies within the atomic range in terms of size. The theory of relativity is the second cornerstone of modern physics and was discovered essentially by Albert Einstein. The theory of relativity is concerned with the essence of space, time and gravity, in other words the microcosmos. The quantum theory and theory of relativity have altered our understanding of nature fundamentally over the past century. Einstein's theory of relativity states, among other things, that there are four dimensions in our universe, that time on Earth passes more slowly than on the moon and that time travel is possible in principle. The theory of relativity is particularly important when examining high speeds[4] and energies, forces of

---

[4] In this context high speeds are speeds which are high compared with the speed of light.

gravity and the effects of time. Quantum physics, on the other hand, confuses us with strange phenomena and predictions which cannot be explained by a conservative body of thought or even intuition. The result of an experiment therefore depends on whether someone is observing or not. Particles fly through impassable walls, cats are dead and alive at the same time, and if this were not enough, quanta circumnavigate the speed of light and communicate in real time information about an unknown link in the universe. Just imagine that your pocket calculator spews out different results for the same calculation, depending on whether you are watching it doing the calculation or not. Quite remarkable, isn't it?

Nature is like a house. The roof is the everyday experience from which we have an intuitive idea of how our world works. The pillars supporting the roof are relativity and quantum physics from which scientists draw their knowledge enabling them to describe the phenomena of our world. If we really want to understand how the world works we must stand back and create a foundation that enables us to find out what is behind the wonderful mechanisms we find in nature. For these pillars are not without foundation themselves. Researchers on all continents are working feverishly to discover this foundation, the combining of quantum and relativity physics in one general theory, a kind of all-embracing theory of everything. They are convinced that one day they will be able to combine both cornerstones of modern physics under one roof, just as was the case with electricity and magnetism in the past. All natural laws could be derived from such a theory of everything. These are laws which are just as applicable in the field of the theory of relativity as in the world of the small, quantum physics. Previously the most prestigious scientists have foundered in their attempts to perform the task of finding a theory which is as valid for describing stars, planets and black holes as for describing small elementary particles. Even Albert Einstein found this a hard nut to crack. For in the micro- and macrocosm different laws appear to apply, pointing to a universe that is far more complex than hitherto assumed. Thus the three spatial dimensions with which we are familiar only represent a small proportion of the hyperdimensional world in which we live. And this knowledge is only the beginning of a long expedition into the depths of the next millennium. A candidate for a fundamental world theory which combines quantum physics and the theory of relativity is just being discovered at many different universities. It could provide the blueprint of our existence, all universes and all dimensions. Every characteristic of the universe could be reconstructed from such a "Theory of Everything", making it possible to recognise and learn how to understand the world, nature and its forces in all their different aspects. However, the greatest of all secrets remains concealed from humanity in the theory of everything.

But let us deal with one thing at a time. We are still a long way from eliciting from nature its second greatest secret. Finally the journey is now underway. Let us flash back and reflect how it has come about that Newton's laws of nature we are still learning about in schools are in fact wrong.

1.3 The demise of the ether

Isaac Newton understood the universe to be a huge clock. Its history follows an unchangeable, predetermined course. Like the pendulum of a grandfather clock, which swings unswervingly from one direction to the other, the universe is a captive of time. Time marks the order of things. It passes everywhere and is always the same for everyone, no matter whether you are sitting on a sofa reading a good book or whether you are pulling up stubborn weeds in your garden. You will probably find reading a more pleasant occupation and in doing so feel that time passes as in flight[5]. In fact time has passed to an equal degree, no matter whether you have been reading for an hour or have been weeding. Isaac Newton considered space and time to be absolute and universal. That space and time are the same everywhere for everyone. His views characterised the world view and welded conventional understanding into a structured, constant universe which is held together by calculable natural laws.

The French Revolution ate its own children, Napoleon conquered Europe and landed up on Elba, a volcanic eruption in Indonesia turned summer into an ice age, Karl Marx mobilised to overcome capitalism, the Vatican ended the inquisition, railways and factories opened up the states, the Swiss built the Gotthard Tunnel and the Parisians laid the foundation stone for the "tragic street light", which unenthusiastic artists later called the Eiffel Tower. Then came the year in which physics embarked upon its last experiment, which was meant to answer the last remaining questions and complete this science. In July 1887 Michelson and Morley tried to prove the ether theory prevailing at that time in one of the most significant experiments in history. According to the theory light can only propagate in the light ether – an invisible substance which fills space, just as sound waves can only propagate in a medium, in air for example. On the moon you can shout as loudly as you like. No-one will hear you. Not even the astronaut standing right next to

---

[5]   In fact time passes in an airplane more quickly than on the earth, as we will see in the chapters on the theory of relativity.

you. The moon has no atmosphere, so no air for the transmission of sound waves. Astronauts can therefore only communicate by radio. The idea behind the Michelson-Morley experiment involved measuring the speed at which the Earth moves through the ether, which is assumed to be motionless. As in an aircraft flying through the clouds (there were of course no aircraft at that time), it was expected that it would be possible to measure a wind in this experiment. The so-called ether wind. Most of the scientific world was convinced that it would be possible to solve the last major problems of physics with this experiment, and some professors even tried to put students off studying physics because "almost everything has been explored in this science and that there are only insignificant gaps still to be filled" [6]. Hardly anyone seriously doubted the ether theory, and if they did they kept it a close secret for fear of losing their reputation and respect. Yet it all turned out completely different. The experiment of Michelson and Morley produced a zero result. The ether was undetectable. Science looked for explanations that would still rescue the ether theory. True to the motto: Although this battle has been lost, the war is far from being so. For on the one hand the experiment was very complicated and extremely susceptible to the slightest outside influences. The traffic around the test facility had even been halted for a short time to cut out vibrations and other disturbing influences. On the other hand numerous alternative theories were circulating, for example the theory that the ether is entrained completely with the earth and that ether wind is therefore not naturally measurable (since in this case

---

[6] This is how the Munich physics professor Philipp von Jolly answered the inquiry of the gifted student Max Planck about the prospects in "physics" as a field of study. It was this very Max Planck who later presented physics with numerous hitherto unanswered questions with the introduction of quantum mechanics.

the Earth and ether moved at the same speed). Yet the demise of the old sciences and views was unavoidable. The tender shoots of research which sprang from the gardens of Newton became a cactus in the park of the new world. However, the most prominent contemporaries had to rely on second guessing him. No one would have suspected thorns where only blossoms could flourish. The Michelson-Morley experiment went down in the history of science like the bursting of a dam. The experiment, which was to have been last in physics, opened the doors to a whole new science. To modern physics. Nature is as it is, whether it suits the scientists or not. No theory can change that. No creative idea is powerful enough to rewrite the rules of nature. But the most amazing thing of all is that beyond the ether theory nature is completely different from how it appears in everyday life. Nature, which was to be discovered in the following years and decades, is radically different from the nature we observe with our own eyes and experience daily. Nature is quite simply so wonderful, fantastic and mysterious because it is designed that way. And not because sensation-seeking authors or audacious scientists wanted it to be so. This is perhaps the most important scientific knowledge gained in the twentieth century.

The Michelson-Morley experiment was to go down in history as the greatest disaster that occurred in the physics of the 19th century. And once again it goes to prove just how little we actually know of reality. For if we accept the assertion that everything of importance has already been investigated, we have to say, looking at the present state of knowledge, that we do not yet know anything. There are so many important questions that remain to be answered and explained. And with each new advance and with each new discovery new avenues of research are opened up. Even today there is no question of this science ever being completed. The ether case marked an historical turning point in the history of physics which was to change science and our world view for ever. The ether experiment paved the way for the discovery and establishment of new theories such as Einstein's theory of relativity, which is still presenting us with numerous unanswered and exciting questions.

2. Einstein and the theory of relativity

Albert Einstein recognised the signs of the times. He was concerned with the inconsistencies emanating from classical mechanics, astronomical observations and the Michelson-Morley experiment. He conceived of a scientific revolution involving the discovery of the fourth dimension, and in his work he abandoned the conservative world view, together with the ether. The special theory of relativity was to shake apart the establish concepts of time and space.

Albert Einstein was a genius. He revolutionised physics and the world view with several theories which won him the Nobel prize. In 1905 he published a treatise on the photoelectric effect which revolutionised how we view light. He also published evidence of the molecular structure of matter, a revolutionary work on the electrodynamics of moved bodies (which is regarded as the core of the Special Theory of Relativity), together with the famous equivalence formula of mass and energy "$e=mc^2$". The unknown Einstein, who had worked in the Swiss Patent Office in Bern since 1902, threw whole world views out of the window in his year of greatness. He discovered substantial effects of quantum physics, attributed puzzling phenomena in the universe to the existence of a fourth dimension (!) and, with the theory of relativity, cemented the second pillar of modern physics. Suddenly physics was again faced with a number of new puzzles. Everything was once again on the table. For years it was believed, incorrectly, that the laws of nature had been largely decoded. Einstein came, saw and conquered. Einstein was always interested in the astronomical dimension of physics. He was concerned with the laws of the macrocosm, the planets and stars, and of the universe. He set out to unlock the secrets of gravity and explain (and not just "how") an apple falls from the tree to the ground. Almost casually he wrote the light quantum hypothesis and made further important contributions to the recent quantum theories which try to describe a set of rules of small-scale physics. However, he did not set great store by many statements of this quantum mechanics, even though he was one of the pioneers and gave the theories the necessary publicity. In particular, he abhorred the idea that quantum motion should be based on coincidences. He was convinced that the elementary particles did not really move randomly and that this assumption would later turn out to be a temporary alibi solution. This conviction led him to his world-famous phrase: "The old man (God) does not throw dice".

Einstein could not go along with the concept of an unpredictable and changing universe all his life. The belief in an ordered structure, in an absolutely predictable world, shaped a goodly proportion of his life. The controversial Cosmological Constant contributed just as much to this as the stable, constant universe he longed for.

The special theory of relativity

Albert Einstein published the special theory of relativity[7] in 1905, his year of greatness. He fundamentally changed our understanding

---

[7] The special theory of relativity describes the essence and relativity of space, time, lengths, masses or energies, but without considering gravity. Only in the general theory of relativity did Albert Einstein succeed in incorporating and explaining gravity.

of space and time. The special theory of relativity establishes two ground-breaking theses. First, it defines the speed of light as the cosmic speed limit. Nothing can move faster than light. No spaceship. No particle. Even if you could tap into the entire universe as an energy source, according to the special theory of relativity, it is not possible to accelerate a mass to the speed of light. Second, the laws of nature are the same for all observers if they are moving uniformly, not accelerating or braking. Accordingly, the same laws of nature apply to the driver in a car on the highway as to an astronaut in a spaceship.

Pardon, Mr. Author, have I misunderstood something here? Surely you are not claiming that the theory of relativity, which is said to have turned half of the world of physics on its head, have begotten only these two half-baked postulates?

Well, in principle, the special theory of relativity is actually based on these two postulates. What reads rather unspectacularly, however, turns into an assault on the trenches of classical physics and hence also on our intuition. These two statements embody a huge potential for conflict with the ideas that we have there. With breathtaking consequences for our world view. For the special theory of relativity is revolutionising our understanding of space and time - and all the phenomena predicted by the theory of relativity really exist. Not just here on paper, not just theoretically in any lab or study, but really, in your world, in my world, in the world in which we live.

But - as usual - one step at a time. In order to understand why clocks in a spaceship go more slowly than in a train, we look briefly back at the mix of general indignation and exuberant irritation that prevailed after the publication of the Theory of Relativity. The special theory of relativity was a nod to the experts, who only now realised how stupidly they had dealt with the accepted physics. After all, it had been believed that the laws of physics were almost completely deciphered. Except the young Einstein. With an actually perfidious thought experiment, he succeeded, to the considerable astonishment of the experts, in changing our world view permanently. Permanently. For even today, very few people know about the fascinating essence of space and time. A very different essence to that we know through our intuition and perception. Einstein addressed a problem that had already failed many renowned scientists. He wondered: What happens if you fly near a ray of light at the speed of light? In this case, the light and the observer[8] move at the same speed, the speed of light. As a result, they rest relative to each other. Just like two trains that run side by side on tracks laid in parallel. The observer can therefore watch the light or reach for it, rather like on a cycle trip reaching for the drinking bottle of your friend who is riding next to you at the same speed but still has water in his bottle. So far, at least, the classical view or pretty much what we might conclude from our everyday experience. The problem is that the Scottish physicist James Clerk Maxwell had already published his "Maxwell's equations" at the Royal Society in 1864. This was an extremely important theory describing the behaviour of electric and magnetic fields as well as their interaction with matter. The point: Maxwell postulated oscillating electric and magnetic fields (now better known as "electromagnetic waves"), all moving at a constant speed. A speed he calculated with the funds then available at around 310,740 kilometres per second. This value was so close to the

---

[8]   In classical physics movement at the speed of light is possible. Only the special theory of relativity recognises that in principle it is not possible to accelerate a mass to the speed of light.

presumed speed of light that Maxwell made the then bold assumption that even light could be an electromagnetic wave. However, electromagnetic waves always move at the same speed in a vacuum, never slow down and never stand still. This was in stark contradiction to the classical assumption that the difference in speed between an observer and the light decreases as the observer moves faster. But if the speed of light depends on one's own movement, the laws of nature are not the same everywhere because the speed of light changes according to the speed of movement and is therefore not constant. Furthermore, there would in this case have to be a preferred reference system, to some extent a perspective of the universe that is preferable to all other perspectives. A perspective in which the laws of nature apply in unchanged form, in which the speed of light, for example, is constant. Such a perspective would be the ether. Physics was therefore faced with a major contradiction: On the one hand Maxwell's promising theory postulated the constancy of the speed of light, and on the other classical physics called for an ether, an absolute reference system, but where the laws of nature do not apply equally everywhere because the speed of light, for example, is slower relative to a fast space ship.

However, could it not also be the fact that Maxwell's theory is incorrect and the classical assumption is correct? Would not the contradiction vanish into thin air?

No, for it can only proved whether a theory is right or wrong by experimentation. A theory which is formulated so beautifully and elegantly on paper may not be recognised as correct if its predictions and theses are not confirmed in reality. Otherwise science would be a collection of arbitrary laws that are meaningless throughout nature. Classical physics is based on the existence of the ether. The Michelson-Morley experiment, however, had shown that the attempt to measure a deviation from the speed of light by the ether wind produced a zero result. Consequently there is no ether and hence no absolute reference system. Accordingly the speed of light would remain the same in every reference system. An important index for Maxwell, who as is well known postulated that electromagnetic waves, and hence also light, always propagate at the speed of light. Einstein took up this contradiction between the classical view and Maxwell's electromagnetism and solved it in the special theory of relativity by making two fundamental statements: the postulate of the constancy of the speed of light and the principle of relativity, according to which all reference systems are equal. Put simply, no matter what you do, light is always moving everywhere at the speed of light, or the laws of nature apply everywhere and always the to the same degree. Some of the most spectacular phenomena of physics should finally emerge from these unsuspected postulates.

2.1.1 The Einstein Postulates

The first postulate of the special theory of relativity states that the speed of light is constant. Light always propagates in a vacuum at the same speed, namely the speed of light. This is about 300,000 kilometres per second, and in one second it covers the distance between the Moon and Earth for which our fastest rockets would take more than ten hours.

What happens when we track a ray of light at the speed of light?

This is where the second postulate comes in: The principle of relativity. It states that two uniformly moved observers are completely equal to each other. So you cannot establish, with any experiment conducted anywhere in the world, whether a train is travelling at a constant pace or is stationary. This has to do with the fact that there is no single absolute reference system, only reference systems that are moved relative to each other. One example illustrates the principle of relativity: you are sitting in an express train and drinking a coffee that the friendly chap in the minibar has just sold you. A commuter is standing on the platform watching the passing train. The locomotive and the wagons, but also the passengers and their coffee are in his view moving very quickly in the direction of travel. The commuter sees your coffee roar through the station at 100 kilometres per hour. From your perspective, of course, the coffee does not move or if it does, then at most from the table to your mouth. If we want to judge this scene objectively, then we are in a dilemma. Or as a judge, how do you assess the claim by the commuter that a coffee rushed through the station at 100 kilometres an hour and the passenger denies this tapping his head with his finger?

The ether theory would argue for the intuition and judge that the train and thus the coffee have moved. The passenger could measure the speed of light in the train and would find that it is less than the speed of light on the stationary platform because the train is of course moving. But as we know, the theory of relativity has filled the breach since it has been experimentally proved that the ether does not exist and the ether theory is therefore false. This is in contrast to the special theory of relativity, which is now regarded as one of the best-tested and confirmed theories of physics. Let us call the special theory of relativity to the witness stand. What can it tell us about the coffee and train?

The narrative is relatively short. According to the principle of relativity, all uniformly moved observers are completely equal to each other. This means: No objective judgment can be given as to whether the coffee is moving or not, even if you perhaps consider the case obvious. The commuter and the passenger judge the scene differently, but both are right in their observations. The fact is that the coffee moves from the perspective of the observer on the platform and rests in the train from the perspective of the passenger. All reference systems are absolutely equal. There is no absolute reference point[9] from which a judgement could be made as to whether this is correct or incorrect. For the laws of nature apply equally in any uniformly moved reference system. Nor is it possible, in principle, to distinguish between a moved and a motionless reference system. So although you feel the jolt when the train leaves, you can also drink your coffee at 300 kilometres per hour, as if you were sitting in a restaurant. The passenger cannot determine whether the train is moving or resting at a constant pace. Now you could argue that he should simply look out of the window, then it could be established very clearly. However, this concept is too short, as I will demonstrate to you later.

The principle of relativity as such is not really revolutionary, since classical physics has more or less emanated from universal laws of nature which apply equally everywhere. What is revolutionary, however, is the inclusion of the constant speed of light. Because that has some very remarkable consequences. Basically, the idea of an absolute world has to be abandoned. Movement, time, lengths, energies or masses are dependent on the viewpoint of the observer. This dependency is what we call "relative".[10] The laws of nature

---

[9] The ether was such an absolute reference point, in other words a preferred reference system from which an objective assessment of the world would have been possible. The Michelson-Morley experiment, however, was the first experiment to show that there is no ether – and that it is not therefore a reference system.

[10] Seven bottles in a wine cellar are relatively few, but seven bottles in a

that apply equally to all of them therefore rarely lead to a situation where the world is exactly the same for everyone. Sometimes the phenomena and processes are much more likely to appear to each observer in an entirely different light. But all approaches are correct and real.

OK. Let's get back to the real problem and focus on an astronaut who boldly dares to attempt to track a ray of light in a spaceship. For this he uses a revolutionary drive that allows the spaceship to travel at speeds close to the speed of light. At the back of the spaceship is a light source that emits a beam of light to the front of the spaceship, where a receiver is installed. From the time it takes for the light beam to cover the distance between the light source and the receiver the astronaut can calculate how fast the light in the spacecraft is moving. If the spaceship is moving at a speed of 100,000 kilometres per second and the speed of light is 300,000 kilometres per second, what speed will the astronaut measure for the light in the spaceship?

Intuitively it might be supposed that the astronaut measures 400,000 kilometres per second as the speed of light, that is the sum of the speed of the spaceship and the speed of light. In the 19th century, although you would not have been congratulated on this result you would have at least attested to a basic knowledge of physics. Until the special theory of relativity was published no-one would have doubted this statement on any scientific basis. Since then, however, things look different. The explanation: the spaceship is a separate reference system. The laws of nature now

---

football team are a relatively large number.

apply equally in all reference systems. A law of nature is the constancy of the speed of light. Light is always approaching at the same speed. The astronaut therefore measures none other than the speed of light as the speed of the beam of light, around 300,000 kilometres per second. From this follows the statement, which is incompatible with classical physics, that it is impossible for an observer in a train to determine whether the train is moving because all experiments provide exactly the same results as when the train is stationary relative to another reference system. Accordingly, it is not possible either to decide which reference system will ultimately move. It is actually meaningless to speak of a dormant reference system as it is moving from the perspective of another reference system and vice versa. Strictly speaking, there are only moved and no resting reference systems. However, to simplify matters, it may still be helpful to regard a reference system as dormant locally, without falsifying the core statement on the situation. Therefore, we will continue to refer to stationary and moved reference systems.

The obvious thing is that the speed of light is independent of the speed of movement of an observer. The speed of light is always the same, no matter how fast you move. If one were to accelerate to close to the speed of light in a hypothetical spacecraft, the light would still be faster by the speed of light (compared to the spaceship). In principle it is impossible to catch up with the light. This was incompatible with the classical view, because in the system of Galileo Galilei the equations for electromagnetic waves, including light, would have had to be adjusted for moving systems. Otherwise the passenger on the train would have been able to measure the speed of the light and thus be able to estimate its speed. The speed of light would therefore have been dependent on the reference system and thus would not be the same for every reference system, which would have required a preferred reference system such as the ether, but which has been refuted by the Michelson-Morley experiment and all subsequent experiments.

This statement is quite irritating. When a train is travelling at 300 kilometres per hour and you are travelling in the direction of travel at 5 kilometres per hour, an observer who is stationary relative to this measures your speed as 305 kilometres per hour. The speed of the train and your walking pace can be easily added together according to classical physics. Now when you enter the cab in the locomotive and turn on the light, what speed does an outside observer measure for the light?

He measures exactly the speed of light, about 300,000 kilometres per second. Even if we transfer the example to a spaceship flying at 100,000 kilometres per second, the light from the headlights still propagates at 300,000 kilometres per second. The speed of light is just about the only thing about the theory of relativity that is absolute. It is always the same, no matter from which viewing point it is measured. It does not matter if it is inside or outside the spaceship. It does not matter if the spaceship flying through the universe or is in the garage at the Space Agency.

Einstein had therefore answered his initial question. It is basically impossible to catch up with light or even overtake it. Light is always travelling at the universal maximum speed, the speed of light. It would be wrong to think that light cannot be overtaken because we do not have the necessary technologies or spacecraft drives. Rather the constancy of the speed of light is of a fundamental nature. Incidentally, this also applies to all phenomena of the theory of relativity or quantum physics. So even with extraterrestrial technology or the skills of a particular civilisation, it would not be possible to catch up with light for thousands of years. Rather the speed of light is a universal law of nature that applies to every transmission of information throughout our universe. In the Stone Age as well as in the time of our great-great-grandchildren.

Often there are rumours about experiments in which information should have been transmitted at supraluminal velocity. You can safely ignore these headlines, as they are usually due to geometric phenomena or idiosyncratic speed interpretations. For example, two spaceships flying in the opposite direction at 80 percent of the speed of light do not move at 160 percent of the speed of light. Rather a relativistic formula must be used for this calculation because the classic speed addition is no longer applicable here, causing the tempo to be constantly below the speed of light. In the chapters on quantum physics we encounter one or two attempts where signals are transmitted at many times the speed of light. However, these transmissions are subject to a not inconsiderable loss of information, thus ensuring that the theory of relativity, in its present interpretation, is not impaired.

2.1.2 Red Wine, Trains and the Pizza Bet

The special theory of relativity leads to a whole host of seemingly peculiar consequences. So you have to completely abandon the conception of an absolute world. The same event seems different to each observer and depends significantly on the reference system used. This applies both to the movement of an object and its location, time, energy, mass or energy. There is no objective reality any more. What's is at the top for you may be at the bottom. What is right can be left. Clocks go faster, sometimes slower. Spaceships are sometimes longer, sometimes shorter. There is no point of view from which we can judge a scene objectively for all observers. Time, space and reality are dependent on the reference system and the movement.

A good example of a bad pizza bet illustrates how significant and far-reaching this knowledge is as far as everyday life is concerned. You stand at the train station and meet an old school friend. You have not seen each other for a long time and decide to look back at old times over coffee and cake. A few hours and two life stories later you realise that it is getting late. To ensure that half a lifetime does not pass until you next meet, you decide to make a small bet, which will be settled the week after. A lunch together at the Italian restaurant. The person whose train leaves the station first has to pay. A fair bet: Both trains have the same departure time.

One week later you meet at the Italian restaurant around the corner to enjoy the agreed culinary delights. You feast on pizza and enjoy a good drop of red wine. Finally, the waitress brings the bill. You put on a friendly smile and decide that your school friend will pay the bill to settle the bet. His train had left before your train. Your school friend looks a bit confused and replies that your train left before his train.

What happened?

The two trains are very modern, which is why they immediately reach their cruising speed on departure without accelerating. Therefore, the two friends sit at the window and watch the other train. Of course, whose train leaves first. Excited of course to see whose train leaves first. A few moments later your school friend sees that your train is moving and that he has won the bet. At the same time you see that your school friend's train is moving and that you have therefore won the bet.

Who is right?

In fact, the bet is undecided. You are both right in your assertion. For what really happened, that is, which train left first, depends on the observer's point of view. Both feel that the other train left first, which is correct from each others' perspective. In this respect, both may claim to have won the bet. Well, of course, you do not want to be beaten that easily. You still have a trump up your sleeve. A witness who has watched the scene on a bridge built over the tracks. The witness confirms - probably not to your liking - that your train departed first. It seems that you lost the bet and you settle your bet as a sporting loser.

Unfortunately, you did not know the special theory of relativity, otherwise you could have reasonably negotiated a draw. In fact, it's just our common sense that tells us that a train moves when someone sees it leaving the station. But the train could just as well say that it did not move and it was the whole world around it that moved. No one could judge who is right because everyone is speaking the truth from their reference system. Since every reference system has equal rights, both observers are right in the statements they make.

At this moment, you are obviously reading this line. You have made yourself comfortable somewhere and are enjoying reading this book in peace. If a low-flying plane crashes next to you or a mosquito is circling around your head, you are intuitively convinced that the plane and the mosquito are moving but you are sitting quietly (at least until it crashes). But if you scale your angle of view you quickly realise that this viewpoint is very relative. So you are most likely somewhere on the earth and are following its rotation. The earth moves around the sun at approximately 30 kilometres per second, corresponding to around 108,000 kilometres per hour. The sun, in turn, revolves quite quickly around the centre of the Milky Way. Are you quite sure that you are at rest?

Of course it is difficult in practice to explain to the police officer that the red light has hit you. The special theory of relativity would certainly agree with you. In fact, we are intuitively stuck to our world view in such a way that the idea that the earth moves relative to the train and not vice versa seems absurd. But if we imagine an empty universe with only two spaceships moving relative to each other, both pilots will feel that the other spaceship is moving relative to their own spaceship. Since there is no absolute reference system in the universe, we basically have no way of telling which spaceship is moving. Rather, the perspectives of both pilots are correct. Again, the evidence is that this a fundamental principle of nature. It is not possible to designate a spaceship that is moving because there is no absolute movement in the universe. A uniform movement always consists of at least two reference systems that move relative to each other. In this case the perception of the movement is relative and therefore dependent on the observer.

Moreover, this effect is used in practically all computer games. When the player manoevres his car through winding hillclimbs, or strolls through the streets as a gangster, he has the feeling that he is moving through the virtual landscapes. Actually computer games are generally programmed so that the entire virtual world moves towards the player. When the player presses the "Accelerate" button his car technically stays put. The street, the landscape and every tree are moved towards the player for this purpose. The player does not consciously notice these perspectives but thinks that his figure or his vehicle is moving along the virtual path. Rather like when we humans intuitively assume that we are moving when we are running through the city and that the world is not moving around us.

Movements, locations or speeds are relative. For example, you might wonder about the trajectory of a ball which you throw vertically into the air in a moving train and catch again? And you will get a most interesting answer. The passengers on the train (and thus in the same reference system as you) will say the ball rises vertically and then falls vertically back into your hands. The trajectory of the ball corresponds to a straight line. However, an observer standing on the platform will perceive the scene quite differently in the passing train. From his point of view, the basic event is the same - that is, you throw the ball and catch it again - but the ball moves differently: The observer on the platform sees the ball flying up in an arc and fall back into your hands. This is because, from his perspective, the train moves and therefore continues even whilst the ball is in the air, so the ball has a longer trajectory back into your hands. As a result, a person on the platform perceives the scene quite differently from a passenger sitting in the train. For the passenger, however, it does not matter whether the train is moving or stationary. The ball always flies and falls in the train (respectively in the reference system). However, both views are justified and correct from each perspective. You could also turn the scene around. If the person juggles with the ball on the platform, the passenger in the train will have the same perception as the person on the platform earlier. From his perspective the person will again feel that the ball is flying and falling exactly vertically.

The same events are perceived differently in space and time by two different observers moving in relation to each other. This leads to strange phenomena such as clocks running slower or fast cars which suddenly fit into garages which are actually far too small. But the principle of relativity applies not only to places, movements, or velocities, but also to lengths, masses, energies, and even time, as we shall see in the following chapters.

2.1.3 Does a second always last a second?

The special theory of relativity goes one step further and also transfers the principle of relativity to time. Therefore time is not a universal quantity that passes equally quickly everywhere and for everyone as classical physics tacitly assumes. Time is much more dependent on the observer and is therefore relative. Each reference system has its own time sequence. In a fast-moving spaceship, clocks go slower than in a car. One minute on your watch does not necessarily mean the same time as one minute in a spaceship. This phenomenon is called time dilation. It is responsible for allowing people to travel hundreds of years into the future without ageing. What sounds incredible has been confirmed in numerous experiments in particle accelerators. Elementary particles such as muons decay more slowly the faster they move.

Of course, it is a matter of opinion whether time passes more slowly or follows its "ordinary" course. Each observer in a uniformly moving reference system does not perceive any time anomaly for the processes in this system. Rather he sees all external events slowed down. When an astronaut looks at his wristwatch in a spaceship, time passes quite normally from his point of view. If he dares to look out of the window, moving reference systems move in slow motion relative to him, and accordingly the clocks slow down. The conclusion that there is more time to live in a very fast-moving spaceship is therefore only partially correct. Whilst it is true that the astronaut in a fast spaceship may, under certain circumstances, survive entire generations and civilisations on Earth, he cannot do more than any other person in his life, as we explain in the chapter on time travel.

But how can this time dilation effect be achieved? What is the explanation for clocks running in reference systems moving relative to them?

A popular example used to illustrate the phenomenon of the relativity of time is that of a light clock. A light clock consists of two opposing mirrors in which light is reflected. Whenever the light beam hits the upper mirror the clock advances by one unit of time. As long as the light clock is at rest, the light is reflected from the upper to the lower mirror and back again along a vertical straight line (shown on the left in the diagram). The path travelled by the light is the distance "d", i.e. the distance between the two mirrors.

The light clock is now brought to 25 percent of the speed of light (on the right in the diagram). Whilst the light is now reflected from one mirror to the other, the two mirrors move forward, from left to right in the figure. As you can see, the light now has to cover a longer distance to get from one mirror to another. Since the speed of light is always the same, but the distance is longer now, it takes longer, from the perspective of an outside observer, for the light to arrive at the upper mirror and for the clock to advance one unit of time. The faster the apparatus moves, the longer will be the distance between the mirrors and the slower the clock and time advance.

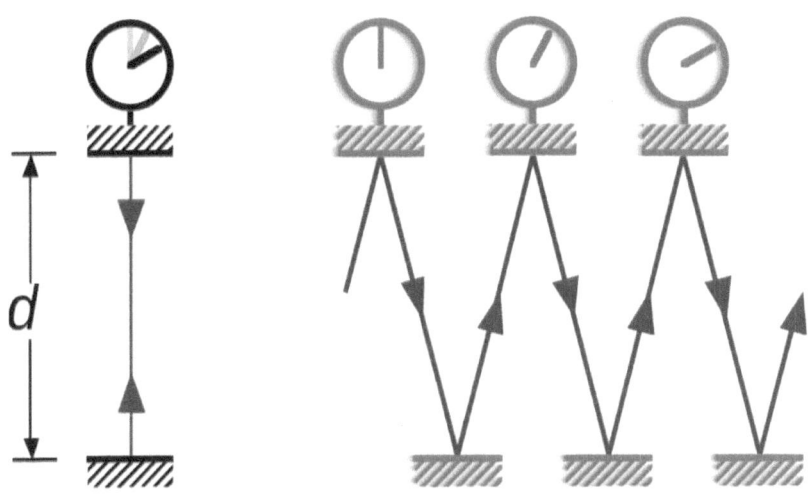

Figure 1: A light clock at 25 percent of the speed of light

This phenomenon is not due to deficient quality or a design feature of the clock used but to a fundamental property of space and time. Space and time are just not absolute, as people once believed and still do believe many places today, but are relative to the observer. The faster a clock moves, the slower a relatively moving observer perceives the passage of time. For the observer within the reference system, however, time passes quite normally. If the light clock is in a spaceship, the astronaut cannot detect a difference over time, even when flying at almost the speed of light. Just like the fellow passengers on the train in the example where a ball is thrown see the ball fly vertically and then fall. To the observers in the reference system, it basically does not matter whether the reference system moves or is at rest. All experiments in the reference system will always produce the same result. An outside observer moving relative to the spaceship sees the clock going more slowly or the ball travelling through a longer trajectory. Incidentally, the perception is interchangeable: when a boy throws the ball straight into the air on the platform, the passengers will see the ball fall in a curve while the train is moving. If an astronaut sees the clock of a satellite running slowly in a fast spacecraft, all the clocks would go in slow motion relative to the satellite in the spacecraft. Incidentally, time dilation does not apply merely to watches. All processes and procedures are subject to time dilation. Even computer calculations, human movements or brain waves are slower.

This example becomes paradoxical when Max, as an outside observer, compares the running of his wristwatch with that of the light clock, which moves at a very high speed. After a while Max realises that 30 minutes have passed on his wristwatch, but on the light clock only 15 minutes have passed. Max concludes that time in the light clock passes more slowly than on his wristwatch. To an observer on board the light clock, this would of course look different. He would again have the feeling that the time in the light clock system was running normally and Max's watch was slow. Max and the light clock occupants are both right in their perception, though they judge time completely differently. But both are right.

Time dilation can no longer be explained by Newtonian physics, but is due to the union of space and time in space-time. Time is not a man-made quantity for introducing a structure into our everyday life. Rather, time is a separate dimension that is inseparably interwoven into the three spatial dimensions we associate with space-time. Our universe consists of four-dimensional space-time. Even now it is certainly difficult for us to imagine the fourth dimension. However, the situation becomes considerably more complicated if we devote ourselves solely to the general theory of relativity. The "special" feature of the special theory of relativity is that its validity is limited to flat areas in space-time, that is, gravity is completely ignored in this theory. In this sense, the special theory of relativity is only an approximation of the general theory of relativity, which also includes curvatures of space-time and with this explains gravity.

Whenever we move through space on foot, by bike or by car, we also move through time. In everyday life, the time dimension remains largely hidden. Calculated in astronomical dimensions, however, the consequences are enormous. For the basic possibility of building a time machine arises for the first time out of time dilation. Before the theory of relativity all speculations about time travel were groundless because no one simply had no idea what time actually was. Since Einstein we have known that time is the fourth dimension. Every movement in space is always associated with a movement in time. Space and time only exist together. If you go into the kitchen to get a cup of coffee and read all the chapters of this book in one go, you will move through space and time. In the general theory of relativity we will see that gravity also substantially influences the course of time. In a separate chapter we will also discuss in detail the possibility of time travel and also tell you about the person who has so far travelled furthest into the future. For time travel is by no means as fictitious as we normally imagine.

2.1.4 Does one metre always measure one metre?

However, the special theory of relativity relativises not only the perception of time but also our idea of space. This is necessary, otherwise our world would be inconsistent. Distances are not generally measured the same by two observers who are in motion relative to each other. For example, the unit of metre with which we determine distances and lengths is a purely arbitrary definition[11], just as two observers moving relative to each other do not agree on the duration of a second. Both watch the other clock or watch go slower and therefore take one second longer. One metre can be perceived quite differently depending on the observer. Scaled up to our everyday life rather disturbing effects are observed. For example, it would be wrong, according to Einstein, to assume that our car does not fit in the garage because it is too short. For our car has no fixed and absolute length or mass. The car is not simply three metres long and weighs 1,000 pounds. For even these familiar characteristics are relative, i.e. depend on the observer and the state of movement. It may be hard to understand, but we have to accept that the absolute and unambiguous world view is a relic of our intuition. And not just on paper. What the theory of relativity tells you is not an invention but the most accurate description in nature we currently have.

Imagine that we are writing in the year 2500 and we are at the national flight test centre. We need to have our flying machine inspected so that it meets all the standards it needs to meet in order to be approved for traffic. At one point, the government decided to ban all flying machines that did not fit into the test commission's

---

[11] At least the original definitions. Meanwhile the metre is measured relative to the speed of light.

standard garage. After the examiner has examined all the paragraphs extensively and has thoroughly examined your flying machine, you are to park it in the garage to see if it meets the new length standards. Unfortunately, your flying machine is about half a metre too long, which means that the two automatic garage doors cannot close. The examiner clearly points out to you that your flying machine will end up in the scrapyard if it does not comply with the law. They consider the situation briefly and decide you want to avoid the scrapyard. You get a few hundred metres run-up with the flying machine, accelerate as if there were no tomorrows and fly full throttle through the garage. The examiner watches the scene from a safe distance, scratches his head in surprise and issues you the licence for your flying machine. Test passed.

What happened? Was the examiner afraid you would blow him up or ruin his standardised garage?

Not at all. The examiner has only seen your flying machine fly into the garage and both gates closing. Consequently your flying machine had to fit into the garage. But how can that be? A flying machine that only fits in the garage when it is flying at high speed but not when it is stationary?

This enigmatic phenomenon is called the Lorentz contraction and is to space what time dilation is to time. A moving object appears shorter to an outside observer than to its occupants. Its length is contracted or shortened. If a metre rule flies through the air very quickly, Max measures only 0.999 metres, 0.9 metres or even less on the ground, depending on the relative speed. Driving a three-metre-long car at 100 kilometres per hour on the highway, the effects of the Lorentz contraction are negligible and with our technical means hardly measurable. In this case, the difference in length is only about 0.0005 millimetres. The difference between this and our classical assumption is just as imperceptible as the time dilation compared to the speed of light at such low speeds. However, if a 20-metre spaceship flies at 99 percent of the speed of light, things look quite different again. Now the spaceship only measures about 2 metres from the perspective of the outside observer. It appears to him to be 90 percent smaller compared to the perspective of an observer who is in the reference system. For the occupants, of course, the size of the spacecraft does not change because they are moving at the speed of the spaceship and are therefore part of this reference system. To the occupants the spaceship appears as if it were standing still (which is due to the principle of relativity). The occupants in turn perceive the world surrounding the spaceship as shortened.

But what happens when the spaceship (purely theoretically) reaches the speed of light?

As we know, it is not possible to accelerate a mass to the speed of light. But we may still wonder what happens if, due to a hitherto unknown anomaly, another dimension or an error in formalism, it could still reach the speed of light. In this case, the length of the spaceship would be zero. The spaceship would no longer have a spatial extent and would therefore lose its spatial dimension to some degree. At the same time, its energy would have to be infinite (otherwise it is not possible for a mass to reach the speed of light in today's understanding). To find out what would happen if it reached even supraluminal speed we give details of this in the chapter on time travel. But if the spaceship flies just below the speed of light, it might shrink to just a few millimetres., making it barely visible. Without, however, the occupants feeling cramped. For according to the principle of relativity nothing changes for them.

As already stated with regard to the other phenomena, it would be wrong to think that the length contraction is an optical illusion causing the outside observer to misjudge the length of the spaceship. Rather is the length contraction as well as the time dilation a fundamental peculiarity of the universe, space-time. There is no absolute right, in principle, when we make statements about the observation of space or time, distances, speeds, simultaneity, lengths, masses or energies.

If Einstein had lived in Newton's time, he would have been bequeathed to the executioner for scalping the absolute world-view. It would never have occurred to him that every person sees the world differently from his own point of view, and yet at the same time goes along with the physical laws of this world. It is merely a melancholic irony that the relativity theory is based on the absolute speed of light. The speed of light is in fact the only thing that is absolute in the theory of relativity. If the speed of light were not constant and absolute, phenomena such as time dilation or length contraction would not exist.

It may astonish you how much nature differs from our perception. But our everyday experience is only an approximation of reality. An approximation that only applies if we assume that the same circumstances apply as in our everyday lives. If we were moving within the range of the speed of light, the world would seem rather strange to us. We would be surrounded by approximations of reality. By and large it is the phenomena of the theory of relativity that remain closed to us in everyday life. On a small scale we do not fare differently, as we will learn in the chapters on the quantum nightmare. We see a chair and no atoms. We see a human and no cells. We "see" light, not a wave. And the edge of the table is square only until we start to analyse this in sufficient detail.

2.1.5 The garage paradox

After Einstein had published the special theory of relativity physicists searched throughout the world for contradictory mental experiments in order to refute them. One of these apparent contributions is the so-called garage paradox[12] which states that your will fit in the kennel if you drive fast enough. This circumstance is due to the Lorentz contraction, a phenomenon of the special theory of relativity whereby a stationary observer sees a fast vehicle shortened. From a certain speed, your car is contracted so much that it at least fits lengthwise in a garage or dwelling, which is far too small for classic scales.

The problem here is this:

We have two different reference systems. On the one hand, the outside observer. He sees the car shortened so that it fits into the garage, which is actually too small. On the other hand we have the car and its occupants. They experience the car in its normal size and instead have the feeling that the distances (the street or even the garage) are shortened. From the perspective of the observer the car fits easily into the garage. From the perspective of the occupants, it is certainly too small, as they perceive the size of the car unchanged but see the length of the garage shortened. What happens when the car drives into the garage? Does it coincide with what confirms the perception of the observer, or does it not

---

[12] Another supposed contradiction is the "twin paradox", which is explained in the chapter "Time Travel".

coincide with what confirms the perception of the occupants?

At first glance, this is a contradiction. After all, the car can fit in the garage or not, in which case a perception will turn out to be a preference [13] This, in turn, would refute the postulate of the principle of relativity whereby every reference system is equal - and with the postulate the special theory of relativity.

Let's play through the scene where the car drives into the garage that is too small and see what happens. Let's imagine a through garage with two doors, one at the entrance and one at the exit. The observer now decides to close the two doors as soon as the car is completely in the garage. He wants to prove that his perspective is right and that the car actually fits into the garage. The car drives at high speed through the front gate into the garage. As soon as the rear of the car has passed the through entrance door, the observer closes both doors and immediately opens them again. The car leaves the garage shortly thereafter without the slightest scratch, proving that it was completely enclosed for a short time. This would prove the perspective of the outside observer. So is he right and does the car actually fit into the garage which is obviously much too small, due to length contraction?

So that we can make a judgment, we look at what happens from the perspective of the occupants when the car drives into the garage. First, the exit door closes then opens again just before the car has arrived there. A collision is therefore prevented. Now the

---

[13] At this point the paradox is not necessarily that the car does not fit into the garage according to the classic view (observed at "rest"), but rather that, from the point of view of the observer based on the length contraction, both perceptions appear to contradict each other. Such a contradiction would in turn be tantamount to a rebuttal of the theory of relativity – if it were to be confirmed and not turn out to be a misunderstanding.

front part of the car passes the exit but the rear part of the car has not yet passed the entrance. The car protrudes from both doors. Finally, the rear of the car passes the entrance. The entrance door closes and opens shortly thereafter. From the perspective of the occupants, the car was therefore at no time completely in the garage, but from the perspective of the observer it is already in the garage. From both perspectives, the scene has played out differently with the same events (e.g. door open, door closed).

But why is the sequence of events in the eyes of the occupants evidently confused compared with that seen by the observer? Why does the rear door close and open again for the occupants before the rear of the car has even passed the entrance door, whilst the observer sees the car entirely in the garage?

The problem is that we presuppose an absolute perception of simultaneity that does not exist in nature. The garage paradox is not a contradiction in this respect but is due to the fact that every observer perceives the simultaneity differently in a different reference system. In reality, however, there is no absolute simultaneity. Two events that took place simultaneously for observer A take place for observer B with a time delay. The consequences of events are the same in all reference systems. If the garage door is destroyed in the exercise, because the car actually does not fit in the garage, it is destroyed from the perspective of all reference systems. Otherwise, the garage door would be at the same time undamaged and destroyed, creating a paradox, as we will find in quantum physics in the case of the Schrödinger cat. The timing of the events may be significantly different[14] This leads to

---

[14] But not the sequence of cause and effect. As long as all the information is transmitted at no more than the speed of light, the causality of the cause and

paradoxically perceived situations.

Going back to our garage example: from the observer's point of view, both doors close and open at the same time. From the point of view of the car occupants, however, the exit door opens and closes first, then the car passes the entrance completely and only then does the entrance door close and open. In both perspectives, the garage and car remain unscathed. The solution to the apparent contradiction is that in relativity the concept of simultaneity must also be understood in relative terms.

2.1.6 Of Zombies, the conductor and simultaneity

The special theory of relativity not only destroys our idea of an absolute space or an absolute time, but also that of universal simultaneity. There is no simultaneity that all observers agree upon. When a train enters the station, the front and rear doors open simultaneously for the conductor standing in the middle of a car. For the passenger waiting at the platform, however, the doors further away open with a delay, as the light from the door further away has to travel a longer distance until it reaches the passers-by. The theory of relativity, however, extends the relativity of simultaneity beyond the pure duration of light. Whenever two events occur simultaneously in a reference system, but not in the same place, there is another reference system in which these events do not occur simultaneously.

---

effect of an event always exists. Only at supraluminal velocity would it be possible for the effect to overtake the cause and for the door to open, for example, before it has been closed.

In one train, a light cannon is mounted at the front and rear. Exactly in the middle is a vampire, who is to be executed, accompanied by the conductor next to him, who records the execution. The train goes very fast through a train station. On the platform there is the responsible judge, who wants to observe the events from a distance and testify. The light cannons fire and the vampire dies. The conductor mentions in the report that both light cannons fired at the same time and hit and eliminated the vampire at the same time. When the judge compares the record with his report, he does not agree. Although both shots have reached the target at the same time, which he agrees with, the rear light cannon fired a few moments earlier.

Let's analyse the special relativity scene to find out what really happened. The two light cannons in the train are in the same frame as the conductor and the vampire. The light beam of the rear and front cannon arrives at the vampire exactly at the same time, because it is in the middle and the speed of light is constant. Both light cannons, viewed from the point of view of the conductor, contributed equally to the elimination of the vampire. The judge on the platform is of a quite different opinion. Seen from his perspective, the rear cannon fired its beam earlier than the front cannon. This perception stems from the fact that the speed of light is constant even in its reference system. Therefore, the light of the nearer cannon (in the example the rear cannon) reaches the target earlier than the light of the front cannon since the train is moving away from the station from the point of view of the onlooker. He will agree, however, that the two rays of light have hit the vampire at the same time, but do not agree with the conductor, that the two rays of light were also fired at the same time. This is because, from the point of view of the observer, the vampire has moved away from the rear cannon due to the high speed of the train, and the light from the rear cannon therefore has to travel a longer distance than that from the front cannon. From the point of view of the vampire who is on the train, the distance between the two cannons has remained unchanged. Instead of relying on the observer's perception, sensors could be placed on the tracks to measure the light beams from the cannon. Classically, the observer could now simply consider the durations of the light (the time it takes the light to reach his eyes from the cannon and let him know that the beam of light has been fired) and agree with the vampire on the time of firing. With the theory of relativity - i.e. as it is to be measured in reality - the judge will also determine, taking into account the light travel time, that the light rays were not fired simultaneously. He will therefore never agree with the conductor, who insists that both beams have been fired at the same time. However, both are right in their observations, which leads us to conclude that the time sequence of events can be judged differently by different observers. The observers will only agree that the event took place. In each reference system, both cannons are fired and the light rays hit the vampire at the same time. It is not possible to decide whether the cannons shoot at the same time or not because that depends on the observer. Speed and location, but also time are very relative here.

For our everyday understanding of simultaneity we can usually ignore this phenomenon. In principle, the relativity of simultaneity could also be measured on the basis of normal characteristics. However, the differences at such low speeds compared to the speed of light are so small that they are virtually impossible to measure.

The relativity of simultaneity can also be interpreted differently. All information in the universe reaches us with a delay because the speed of transmission is limited to the speed of light. This applies not only to photons or signals but also to gravity. In the centre of the solar system there is, as we all know, our sun, which attracts the surrounding planets by its gravity, keeping them in a stable orbit. If the sun were to disappears at this moment it would take about eight and a half minutes for the earth to fly out of orbit. This is the time that gravity takes to transmit its power from the sun to the earth. The planets of the solar system are not thrown out of orbit at the same time. Nearer planets like Mercury or Venus would be affected first, followed by Earth, and finally, the dwarf planet Pluto, which is farthest from the Sun. Since information can propagate at the maximum speed of that of light, our view of the present is basically quite limited. If someone from the centre of the Milky Way were to directs a telescope with very high resolution towards the Earth, he would see civilisation as it was about 30,000 years ago. No cars. No cities. No pyramids. Likewise, the reception of an extraterrestrial signal would for a long time give an indication of extraterrestrial life. Such a signal could mean that there was once life on another planet. In view of the astronomical distances that the light had to travel to the earth, it would not be impossible for this civilisation to have already perished in the meantime.

2.1.7 Meteorites, astronauts and mass

In the previous chapters we have discussed various phenomena of the special theory of relativity that are as unexpected as they are astonishing. Time dilation makes clocks go slower, the length contraction allows fast cars to be parked in garages that are to small. But that's not all. One crucial element is still missing.

A meteorite is flying towards the moon. The space agency on Earth fears that the impact will have devastating effects on the orbit of the moon. In order to track the trajectory and determine the danger more accurately, a spaceship is launched which moves at approximately the speed of light relative to the meteorite. The astronaut analyses the situation and then transmits to Earth the information that the meteorite is no longer dangerous. It is much slower and smaller than initially thought. Shortly thereafter, the meteorite hits the moon with a huge impact that throws it out of orbit, leading to tremendous floods and monster waves on Earth. The astronaut does not believe his ears when he is told that he must have made a mistake. He saw exactly how small and slow the meteorite was. It should not have caused any significant damage.

In fact, one important aspect of the theory of relativity was forgotten. Otherwise a warning would have been given. For astronaut saw the meteorite reduced by the time dilation and Lorentz contraction and perceived that is was moving much slower than it actually was as viewed from the Earth. But the event must always be the same from every perspective. So how is it possible for a slow, small meteorite to develop such a destructive force that the moon is blown out of orbit?

An important aspect of the theory of relativity is the mass increase. The mass of a body is by no means constant, as we assume in classical physics. Rather mass is associated with the speed at which a body moves. The more massive it is, the faster it is. This is also why no body is able to reach the speed of light. For For to do this it would have to mobilise an infinite amount of energy and would thereby collapse into a black hole - so far as our current knowledge tells us. The theory of relativity also states that mass and energy are mutually convertible and hence thus in principle comparable. Driving a car at ten kilometres an hour into a concrete wall will limit the damage to the wall. If you drive into the wall at fifty kilometres an hour, however, the car or the wall will be demolished, depending on design and structure. But the same amount of damage can be caused by a car driven at ten kilometres an hour as one driven at fifty kilometres an hour kilometres per hour by increasing the mass of the car. For example, if a truck weighs several times the weight of a car driven at ten kilometres per hour into the concrete wall, the damage will also be considerable. This knowledge may also explain the apparent contradiction between the assessment by the astronaut that the meteorite was harmless and the actual force of the impact. The meteorite seemed to the astronaut to have been reduced in size and slowed, but he should have noticed that its mass was considerable. The faster a body moves, the more mass it has. Due to the increased mass, the enormous impact can also be explained from the astronaut's point of view, thus the theory is self-contained and consistent. For we know that the events themselves are equally devastating in all reference systems. From an astronaut's point of view, the impact of the meteorite must ultimately have had the same devastating effect as from the perspective of the space agency observing the event with a telescope. The meteorite, however, had more mass than perceived by the astronaut is he was at a greater speed than the meteorite. From the astronaut's point of view, the higher mass compensated for the slower crash velocity of the meteorite associated with time dilation, just as you can cause the same amount damage with a truck driven at a slower speed than a sports car at high speed. The relativistic mass increase should not be imagined as a conventional mass that can be measured with a balance in a gravitational field. Rather, it concerns kinetic energy, which may also be understood as a mass. For mass and energy are in principle the same and are mutually convertible, as expressed by

Einstein in the special theory of relativity in the formula "e=mc²".

2.1.8 Einstein's formula "E = mc2"

It is arguably the most famous formula in the world, and when it comes to exploring the construction of the first atomic bomb at the regulars' table in the pub it is often cited to castigate Einstein as its inventor. In other circles people are also always happy to declare that this formula ushered in the atomic age and decided the Second World War. This formula is on everyone's lips, and it is astonishing how much a mathematical equation can easily trip off the tongue when it is linked to a dark legend and a couple of emotions. In fact, hardly anyone knows what lies behind this formula, let alone what the individual letters mean. They do not encrypt a blueprint for a nuclear weapon, otherwise the efforts of many nations to upgrade would certainly not have been lost for decades. On the surface the formula is even more closely related to an antimatter bomb than to a nuclear weapon. A weapon of mass destruction, which today can really only be found in Dan Brown's books and can hopefully never be built[15].

But what does $E = mc^2$ actually mean?

First, this equation states that matter and energy are based on the same principle. They are equivalent to each other. In other words, energy and matter are fundamentally the same thing. Nature knows no essential difference between a fire and a stone. Well, you cannot fry sausages with a stone and do not slice them with a fire. We

---

[15] Because of this we have not even tried to build it. History teaches us that we almost always do what we are able to without thinking about whether we should actually do it.

agree on that. But the formula also states that a stone can in principle be converted to a fire and a fire into a stone. And the formula indicates how much energy a stationary stone contains and how much mass a fire contains.

You may be amazed by the fundamental equality of energy and mass. Maybe not. In any case, these three letters changed the world. Perhaps you were even the cornerstone of the theory of everything, trying to grasp all natural laws and combine them in one equation. The idea that the Earth, the moon, the solar system, the ISS, the paper in this book page, your hands, any pen, the air, yes, simply everything that is made up of a few simple elements, is fascinating. Who would have thought that a fire and a stone have so much in common? Secondly, Einstein's equation accurately calculates how much energy can be gained from a given mass or how much energy is needed to produce a given mass. The big "E" stands for energy, the small "m" for mass and the small "c" for the speed of light. The equals sign assigns to energy the product of mass and the speed of light squared. Energy is equal to mass times the speed of light squared. From this formula it may be deduced a very large amount of energy can be obtained from a small mass, such as a pebble. Conversely, it takes a lot of energy to materialise a pebble. The perfect opportunity to buy a quarry and to hoard as much debris and gravel in the basement as possible with a view to becoming a serious competitor as an energy tycoon? Of course, in practice it is not that easy to turn a stone into pure energy. Otherwise nuclear power plants and petrol refineries would have long become obsolete and we could satisfy our energy needs from waste and garbage. As Einstein's formula and later numerous experiments prove, however, this is possible, at least in principle. It is possible. But how do we turn a cold stone into a high-energy fire? In principle, how can we turn any form of matter into energy? And how do we materialise energy?

There is an abundance of sledgehammer methods. For example, you can book a ticket to the ISS at the Russian Space Agency (cost: about $ 20 million) and throw a stone from there towards Earth. If you are not launching a satellite (cost: about $ 500 million), the stone falls into the ozone layer and burns up. The stone is turned to energy in the form of heat.

Mind you, this variant is not practicable when it comes to producing energy. Moreover, we do not necessarily achieve the desired effect in this example, namely that of completely turning the stone into energy. Not to mention the big wallet you need to be able to throw the stone at all. If you've already read the chapter on antimatter you may perhaps see the light when reading these lines. That's because we've discovered that matter turns completely into energy when it collides with antimatter. Einstein's formula gives us nothing but the amount of energy that is released in such a reaction. And those amounts of energy are huge. Or would you have imagined that there is more energy in a stone than in a nuclear bomb?

For example, a 5kg dumbbell releases the energy of about 1000 Hiroshima atomic bombs when converted into pure energy. This could provide all households in Germany with energy for almost a year. A drop of antimatter is more than just a drop in the ocean. Although we are not yet able to use matter in this form as a resource, these examples should show you the practical meaning of $e = mc^2$.

What does this formula, which has so many puzzles, really have to do with the atomic bomb? Actually nothing at all. Or at least not much. Nuclear fission converts only a small fraction of the fissile material into energy. Einstein's formula is therefore not very useful for calculating the energy yield of a nuclear weapon. Nor does e=mc² explain how to build an atomic bomb. Basically, the formula tells us something about nuclear fusion rather than nuclear fission. It would therefore be a big mistake to cite this famous formula as the main prosecution witness against Einstein. Firstly, there is no causal link between the atomic bomb and this formula. Secondly no researcher who does not serve the cause of war can be reproached for discoveries that are later unintentionally misappropriated. Otherwise you would have to take Newton critically to task and blame him for plane crashes and a falling apple. Or blame Tesla for the electrifying effect of power lines. Thirdly this formula construct, which has been attributed to Einstein for decades, did not develop on its own. In fact, Einstein did not discover $e = mc^2$ at all. Before that, various physicists, including Poincaré in 1900, had already derived this formula. However, they never succeeded in incorporating it in a more comprehensive theory and explaining it. In modern physics, the interpreter often earns his laurels. The composer of the formula remains completely unknown.

Although the formula has nothing to do with the atomic bomb or has at least not contributed to its construction, Einstein is not entirely innocent in this either. There are some stains on his character. It was this very Einstein who, in August 1939, had asked the then acting US President Roosevelt in a letter for the atomic bomb to be built. This eventually resulted in the Manhattan Project, probably the largest and most mysterious military research project ever undertaken, involving tens of thousands of people, most of them having no knowledge at all of the overall purpose of their work over the the years. From the cleaning lady to the elite scientist, every man and his dog were brought together in a specially built secret city known as "Site Y" near Los Alamos. The result destroyed Hiroshima and Nagasaki in August 1945 and led the US and Soviet Union into the Cold War. Einstein wrote the letter, however, at the urging of a certain Leo Szilard, who feared that the German nuclear research might bear fruit and give Hitler an atomic bomb. A pre-emptive strike by the Nazis had to be prevented at all costs. That was the reason why Einstein deviated from his pacifist line and moved Roosevelt to action, and not because of the potential for annihilation or because of dubious military scientific importance. Einstein's historical influence on US nuclear efforts was therefore more political than scientific.

2.2 The General Theory of Relativity

Albert Einstein published the general theory of relativity in 1916, thereby crowning his scientific career. It forms the basis of the exploration and understanding of our universe. The general theory of relativity extends the special theory of relativity to gravity and explains for the first time what lies behind this mysterious force of nature that allows apples to fall on heads. According to Einstein, gravity is not an intrinsic property of masses, as Newton supposed. Rather space-time is curved by the presence of masses and energies, creating gravitational force. These curvatures in turn influence the course of events, the flow of time or the weight of a body.

The principle of equivalence forms the conceptual basis of the general theory of relativity, which predicts the basic equality of acceleration and gravity. Just as in the special theory of relativity it is not possible with any experiment to distinguish a stationary from a uniformly moving reference system, in the general theory of relativity an accelerated reference system cannot be distinguished from a reference system in a gravitational field. Thus, it is not possible to determine from a rocket falling on the earth whether the cause of the crash is its own activated engine or the force of attraction of the earth. The course of time, in turn, depends on the extent and strength of an acceleration or gravitational field. A clock on the sun is slower than on earth. On the event horizon of a black hole the clock actually stops because of the enormous space-time curvature.

The General Theory of Relativity is the most accurate description we have of the universe at present. It has been reviewed and confirmed several times over the past decades. The breakthrough came on the occasion of a solar eclipse on May 29, 1919, when British astrophysicist Arthur Eddington observed that the deflection of light by the sun was closer to Einstein's prediction than Newton's classical notion. Einstein, with his astounding theory of space-time, had won the race against classical physics in a first round and became an acclaimed pop star of science overnight. He therefore became the epitome of genius, the embodiment of genius and giftedness and from then on became a man in demand. For the theory of relativity Albert Einstein never received the Nobel Prize, the most coveted award in physics in general, even though well-known research colleagues such as Max Planck had repeatedly nominated him for the award. The committee was afraid that the theory of relativity might turn out to be wrong in retrospect. So Albert Einstein was awarded the Nobel Prize in November 1922[16] for the photoelectric effect, a theory that

---

[16] He received the Nobel Prize for 1921 but the award ceremony did not

contributed significantly to the development of quantum mechanics.

2.2.1 The curved space-time in four dimensions

The general theory of relativity deals with the question of what gravity really is, why all masses attract each other and the connection between gravity and space-time.

Gravity has long been a seven-seal book for science[17]. Although in the 17th century Isaac Newton formulated the first laws of gravity that allowed the falling apple to be calculated as practically as the orbits of the planets, no-one was able to explain in a scientifically sound manner how the apple falls to the ground, why the Earth revolves around the sun or where gravity comes from. Nobody knew what was behind gravity. Nobody knew why two masses always attract and never repel. Newton had no explanation for it either. The decoding of one of the last great secrets of the planetary orbits of our solar system should also lead us beyond Einstein's world view. It has been known for some time that the planets do not revolve around the sun in a perfectly elliptical orbit, but that there are mysterious orbital deviations. Although Newton had already considered two hundred years ago that planets could also attract each other, thereby disrupting elliptical harmony, there was an inexplicable deviation between mathematical theory and astronomical practice. In planets close to the sun, such as Mercury, the error was much larger than was the case with more distant stars.

---

take place until November 1922.

[17] And is still posing numerous puzzles...

The Strange Universe: Einstein, Quantum Physics and the ToE

The French astronomer Urbain Le Verrier prophesied in 1859 the existence of an undiscovered planet "Vulcan" between Mercury and the sun. Numerous astronomers searched with their telescopes for the mysterious heavenly body, a complicated and dangerous venture. On the one hand the sun outshines any nearby celestial body because of its brightness, and on the other hand looking into sunlight can cause visual damage and blindness if the telescope does not work properly. Despite the efforts, "Vulcan" could never be found at the predicted location. Only the general theory of relativity would throw some light on these inconsistencies. It extends the special theory of relativity, which only applies in flat space-time, that is, only when there are no or negligible gravitational influences. For this reason it was later given the name "special" because it only fully applies if gravity is negligible.

The theory of relativity combines space and time inseparably in four-dimensional space-time. The entire universe is therefore a fabric of space and time. Whenever you move through space you also move in time. Here time is the fourth dimension next to the three dimensions of space, the fourth dimension. The general theory of relativity explains gravity with the geometric structure of this space-time, and all the masses and energies bend space-time. These curvatures cause the force of nature that we perceive as gravity. Accordingly, gravity is not a property that proceeds directly from masses but is due to a curvature of space-time caused by the presence of masses [18]. Two bodies always attract each other, as Newton rightly stated. This phenomenon is by no means magic or witchcraft but is attributable to fundamental geometric principles of space-time. Gravity is therefore a fictitious force caused by the curvatures of space-time. A mass, the sun for example, bends space-time, causing a sink in space-time into which close bodies

---

[18] Masses and energies bend space time. According to the special theory of relativity masses and energy are equivalent.

slip due to simple geometrical factors. In everyday language, the bodies are attracted to the sun in this case. A curvature of four-dimensional space-time can hardly be represented in images19. Figure 2 shows how the space time curvature passing through the Earth can be represented diagramatically. In this case the satellite circles the Earth along the curvature.

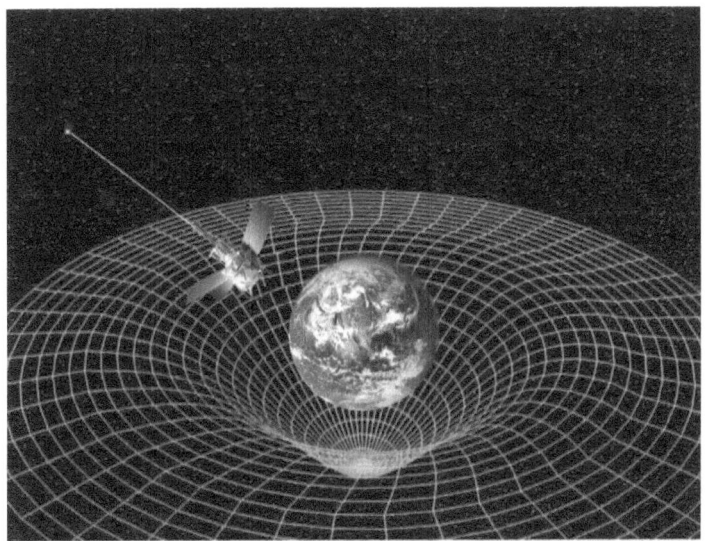

Figure 2 Space-time curvature of the Earth

Our mind is only designed for three spatial dimensions and even then it is difficult for us to visualize the curvature. However, the following example is intended to give you a rough idea of a space-time curvature.

Imagine a piece of aluminium foil. You hold the left end with your left hand the right end with your right hand so that the film is relatively slack. This slide symbolises space-time. As long as you keep the film taut it corresponds to a flat space-time, as required by

---

19   In Figure 2 this fact is simplified to a two-dimensional representation of the curvature.

the special theory of relativity, so that its formulae apply without restriction. There is no gravity in a flat space-time. Now you visualise an orange (fruit). The orange symbolises a heavy mass, let's say the sun (which in terms of colour and shape is an approximate fit). Place the orange in the centre of the film, then watch the film bend downwards unless excessively stretched, to create a sink. If you now take a peanut and place it somewhere in the sink, the peanut will slip down into sink towards the orange. This slide corresponds to the space-time of gravity transmitted by the mass (for example, the sun). Each mass in space-time causes dips and bends into which other masses slide, giving the impression that the masses are attracting each other. The larger a mass, the steeper the resultant dip, and correspondingly the higher the acceleration experienced by a body approaching that sink. In other words, the larger a mass the stronger the attraction that emanates from this mass. The sun, with its huge mass, causes a strong curvature in space-time. The Earth, which "weighs" much less, also causes a curvature, but it is much smaller. Gradually the blue planet now slides down the space-time curvature ("sink") until it reaches the centre (= sun). In reality it is thanks to the centrifugal orbital motion of the Earth around the sun that the blue planet has not yet fallen down into the star - as the peanut in the example would do with the orange. It is important to note that in space-time the sun is not pulled down by the force of attraction, thereby causing a sink in the space-time fabric, as the orange does on the aluminium foil. The space-time curvature, in the form of the sink, has the greater force of attraction, i.e. gravity. Mass bends space-time and space-time causes curvature an acceleration of other masses due to this curvature. This acceleration is what we call the force of attraction or gravity. An invisible hand, which, as in Newton's day, transmits the power of attraction from one body to another, no longer exists in general relativity. Gravity is solely due to the distortions and curvatures of space-time. As you can see, gravity is by no means an inexplicable phenomenon but is elegantly derived from the geometry of space-time.

Incidentally, the radius of gravity is infinite in the general relativity theory. The gravitational effects of our sun would therefore be measurable at the very edge of the universe. At least theoretically. However, it should be considered that gravity is also subject to the speed of light. In the centre of the Milky Way is a black hole whose gravity will have been on its way for some 27,000 years before it reaches the earth. In principle, however, the range of gravity is infinite, according to today's lack of understanding. The gravitational force of our sun reaches every nook and cranny in the universe if there is sufficient time for it to propagate that far. It is quite possible that a future quantum gravity theory will limit the range of gravity. In such a theory, gravity would be quantised like the other natural forces, that is, its effect could only be transmitted as a multiple of a very small amount of gravity. Just as you would pay with only the smallest coin of a currency, for example the 1-cent coin in the Euro currency, but this cannot break down smaller than this. If gravity is transmitted as a multiple of a very small amount of gravity, the gravitational effect cannot be arbitrarily smaller with increasing distance since the smallest packages of gravity cannot be broken up. Consequently, the range of gravity would not be infinite, as the general theory of relativity assumes, but would dissolve at a certain distance. However, the quantisation of gravity has been unsuccessful to date and, together with the union of the three non-gravitational natural forces, represents one of the greatest challenges in physics. Such a unifying theory could also provide the key to unresolved questions such as the gravitational constant. The gravitational constant defines the extent to which the presence of masses and energies bends space-time, thus giving rise to a gravitational force. The gravitational constant is the natural constant with the greatest inaccuracy since it is very difficult to measure precisely. In principle, however, natural constants are also an indication of the incompleteness of a physical theory. Once we have really understood the structure of space and time and the cause of the forces of nature, the natural constants could arise directly from the formalism of such a theory. The gravitational constant could in this case be closely associated with the microscopic structure of space and time. In some theories, such as loop quantum gravity, it looks as if space and time would also be quantised, giving it the smallest unit of time and space. The string theory in turn assumes the transmission particles of gravity to be annular threads, which is why they are able to penetrate all dimensions. However, this is merely conjecture thus far, in contrast to the general theory of relativity, which is experimentally very broadly supported and will remain at the core even in a possibly higher-level theory, just as the Newton laws have maintained their validity as a limiting case of the theory of relativity for low velocities and energies.

You may have wondered why the earth does not fall into the sun, even though the sun causes much greater gravity than our home planet. As we have briefly mentioned, it is due to the movement of the sun that we are in a relatively even orbit. Earth follows in its orbit the principle of least resistance.

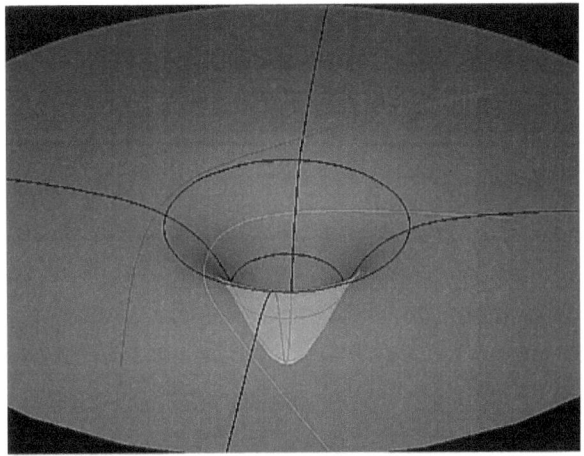

Figure 3   The shortest distance from A to B

The same principle is followed by all other natural processes. No river flows uphill, no beam of light flies through the universe in a zigzag course. The same applies to all celestial bodies and planets. On a flat space-time (in the absence of gravity) this path would be a straight line. From A to B, the shortest route according to classical perception, is simply the direct route. On a football field, the shortest path from the left to the right runs along the centreline and is therefore a straight line.

However, in the curved space-time, that is the universe in which we live, the shortest path no longer runs directly from A to B, but along a geodesic. Consequently, the shortest path in space-time does not follow a straight line but runs along the space-time curvature, as shown in Figure 3. In fact, a straight line in four-dimensional space-time is the longest possible path to follow from a starting point to a destination. This is because space and time are inextricably interwoven. A movement in space is therefore always accompanied by a movement in time. A curvature of space-time always gives rise to the strongest gravity in the centre and weakens with increasing distance. If we try to move from the left side of the curvature to the right side of the curvature in the figure, the straight line would pass directly through the centre - and thus through the area where gravity is strongest. The straight line represents only intuitively the shortest path, but we forget that space and time are inseparably interwoven and therefore the dimension of time must be taken into account. Because gravity causes a slowing of time and runs straight through areas of higher gravity, the straight line turns out to be the slowest of all possible direct paths. Of course, this effect is insignificant in everyday life since the gravitational forces created and the distances are too small to be able to measure a difference between a geodesic and a straight line. In astronomical observations, however, the geodesic is seen as a trajectory of light, for example, because the stars behind the sun become visible from the earth.

Isaac Newton had already observed that every object is attracted by the earth at the same speed. An apple falls to the ground at the same speed as a feather (at the same initial height and in a vacuum). In space-time, the fall of an object due to gravity is nothing but an acceleration of the object along the shortest path, that is along the geodesic. Strictly speaking, the apple does not fall along a straight line from the tree but along a geodesic. The gravitational acceleration is independent of the body since it is only due to the geometry of space. This postulate was tested by the crew of the Apollo 15 mission on the moon with a hammer and a spring. Although the quality of the video was rather poor and correspondingly not so easily recognisable from the experiment, the hammer and spring seemed to hit the lunar surface at the same time. The same experiment would normally fail on Earth because air resistance brakes the spring more than the hammer.

The general theory of relativity explains gravity in terms of curvatures of space-time, which are caused by masses. Other masses slip into these bends, causing gravitational acceleration. The shortest path in a curved space-time is a geodesic and not a straight line, since space and time are inseparably interwoven and therefore every movement in space also means a movement in time.

In the following we will focus more intensively on the passage of time and find an answer to the question of why time in a gravitational field passes more slowly. How is it possible for time to stand still on the event horizon of a black hole or to go slower on the sun than on Earth?

2.2.2 The passage of time and gravity

In the special theory of relativity we have come to know peculiar phenomena such as time dilation or length contraction. Phenomena that show the relativity of space and time in an impressive way. Surely you are wondering to what extent the general theory of relativity predicts similarly amazing time effects. To answer this question, let us dare to make a brief comparison of the two theories. The special theory of relativity is content with considering regular movements and elegantly ruling out accelerations. And hence also gravity, since acceleration and the effect of gravity cannot be physically differentiated according to the equivalence principle. The general theory of relativity is a further development of the special theory of relativity, in which it takes accelerations into account and explains gravity through distortions and curvatures of space-time. However, all phenomena from the special theory of relativity are still valid. Moving objects appear shortened to a stationary observer and watches go slower in fast spaceships. In the special theory of relativity, the Lorentz contraction or time dilation is based on the observation of an event from a certain perspective. Max sees a spaceship shortened by 25 percent because it is moving at high speed. Likewise, Max has the feeling that the spaceship is moving in slow motion, time passing by more slowly. The occupants of the spaceship come to the opposite conclusion. Max seems "shortened" to you and to live in slow motion. Both reference systems make the same statement through the other reference system, but they cannot agree on who is right. Within the special theory of relativity Einstein considers only reference systems that are in regular and linear motion. The general theory of relativity, however, takes one decisive step further. It also considers accelerated movements that cause a situation where symmetry is no longer guaranteed. This will not allow symmetric relativistic effects. Effects that can be used to produce asymmetrical time phenomena, such as time travel. One such asymmetrical effect is gravitational time dilation, which states that time decays increasingly slowly as the strength of a gravitational field increases. In other words, the stronger the gravity the slower time passes. On Earth, time passes more slowly than on the ISS space station. Unlike time dilation, this effect is not symmetrical, which means that for the astronauts on the ISS time on Earth passes more slowly, by a few billionths of a second per year. To an observer on Earth, however, the time experienced by the astronauts on the space station passes more quickly. This means that the astronauts and the observer on Earth agree on how time runs, so they perceive the different time sequences in the same way - unlike in the special theory of relativity, where both perceived time dilation from their own perspective. With time dilation, which is due to rapid movements and not to gravity, the astronauts and the observers on Earth both have the impression that the other relevant reference system is running in slow motion, so their perception is symmetrical or mirror image. Both have the feeling that the other person is moving in slow motion, during which both agree as follows in regard to the gravitational time dilation: Time passes more slowly for the observer in the strong gravitational field. This results in a significant difference between the time dilation of the Special Theory of Relativity and the gravitational time dilation of the General Theory of Relativity: gravitational time dilation produces a sustainable effect, whereas the time dilation of uniform motion has only one effect as long as the reference systems are in motion relative to each other. This circumstance can be illustrated by a simple example: On the event horizon of a black hole the space-time curvature is so strong that not even the light of the gravitational force can escape. Max is a few kilometres ahead of the event horizon. The space-time curvature is already so strong that an outsider has the feeling Max lives in extreme slow motion. Max, on the other hand, would feel that time is passing at an extreme rate for this outward observer. This phenomenon of gravitational time dilation is not only a relative phenomenon but also displays an "effective" effect. For Max time passes very slowly from the observer's point of view. And this is true. If the observer has spent 10 years on Earth, from his point of view, Max will only have aged a few days during this time. In this respect, the gravitational time dilation differs from the time dilation of the uniform movements. For the latter only has an effect as long as two reference systems are in relative motion. When Peter and Hans move for five years at a very fast speed in a flat space-time (a space-time without gravity), they repeatedly notice that each other's clock is slower. Both have the feeling that the other lives in slow motion. If both of them suddenly stopped, Hans and Peter would realise that time has passed for both of them. Hans and Peter would both have aged five years. However, this only applies as long as Hans and Peter are moving at a constant pace and in a straight line. For example, as soon as one changes direction, time dilation has an irreversible, lasting effect. In contrast, the effect of gravitational time dilation is always permanent and not mirror-image. In a strong gravitational field, time actually runs more slowly than on the outside. A person on the edge of a black hole ages much slower than a person on Earth. Time passes on a peak faster than at sea. In the GPS navigation system, therefore, a frequency correction of the signals is performed to avoid positioning errors (even though the effect of gravitational time dilation in terrestrial spheres is relatively small).

But why does time pass more slowly in a gravitational field than in flat space-time? How is it that clocks are slower on Earth than on the Moon?

Imagine a rocket. If the rocket moves at a steady speed, according to the special theory of relativity, it cannot be determined from the rocket whether it is stationary or flying at a constant speed through space. Due to the constancy of the speed of light, time passes normally from the perspective of the occupants. But when the astronauts look out the window, they feel that time in relative frames is slower. The observers from this reference system again have the feeling that time in the rocket is lost in slow motion. To illustrate, the rocket has a light clock like the one we described in the Special Relativity chapter. If the rocket is accelerating now [20], the light beam has to travel a longer distance to get from the upper

---

[20] Acceleration and gravity are physically indistinguishable according to the equivalence principle. Therefore *every* accelera*tion* has the effect of slowing down time.

to the lower mirror and back to the upper mirror from the perspective of an outside observer and the occupants of the spaceship for the simple reason that the mirrors move in the direction of acceleration of the rocket. Therefore the light has to cover a greater distance. Since the speed of light is always the same, the light needs more time to cover the longer distance, which makes time in the spaceship go more slowly. Of course, this applies not only to the light clock but also to all processes in the spaceship, including the onboard electronics or astronauts. If the astronauts take a look out of the spaceship they would have the feeling that everything around the spaceship is moving at a fast speed. An outsider, on the other hand, sees time in the spaceship go more slowly. The gravitational time dilation thus causes an effective change in the time sequence, which is perceived in the same way by each observer. The astronauts could synchronise two atomic clocks before departing from Earth. One atomic clock they deposit in ground control, the other atomic clock is placed in a satellite in orbit. After some time, both atomic clocks will be compared in ground control. You will find that the satellite's atomic clock is on and the atomic clock is under ground control, so that time on Earth has been effectively slower than on the satellite. Evidence of gravitational time dilation. Incidentally, such an experiment was actually carried out by placing two synchronised atomic clocks at different heights. One on a peak top, the other in the lowlands. After a few weeks it became clear that time in the lowland had passed slightly slower than on the peak top because gravity is stronger in the lowlands than on the peak top. To a resident of an island state at sea level time actually passes more slowly than to a hermit who lives in a cave in the Siberian peaks. However, the difference is so small that it cannot be measured with normal wristwatches. However, even with GPS satellites these time effects must be considered. The gravitational time dilation is not an invention of witty scientists but a peculiarity of nature. Time actually passes on the beach more slowly than on Mount Everest.

Moreover, the general theory of relativity has been confirmed experimentally several times. The solar eclipse of 1919 had already broadened the space-time curvature in daylight. Arthur Stanley Eddington found that the position of some stars had apparently shifted. Heavenly bodies which were supposed to be behind the sun revealed themselves in the sky above our native Earth. Through the space-time curvature the light from the stars is directed around the sun so that the stars that are actually invisible also become visible from the Earth. The same phenomenon is confirmed in the observation of distant galaxies. Light, which originates from heavenly bodies, follows a path to the left and right along the space-time curvature of the galaxy. It follows the shortest path that leads according to the general theory of relativity along a geodesic and thus around the galaxy. This results, for the earthly observer, in the illusion of two light sources to the left and right of the galaxy. Both light sources, however, emanate from the same same object. This phenomenon is known as the Einstein Cross. The light is deflected by the curvature and thus strongly confirms Einstein's postulated structure of space and time. In this way, it is possible to observe celestial bodies which are unlikely intuitively to be visually perceptible.

2.2.3 The Shadows of Gravity

As early as the 17th century Newton succeeded in incorporating gravity in a mathematical formula, thus embodying this fundamental force of nature in a calculable framework. However,

Newton did not know what was behind gravity or how gravity works on the inside. Albert Einstein was already able to explain the essence of gravity with his general theory of relativity much more clearly and trace it back to a universal origin. As a result, curvatures and distortions in four-dimensional space-time are responsible for the force we perceive as gravity.

Brian Green, Lisa Randall and many other physicists are racking their brains over the next step. The mystery of gravity is far from being decoded because the force of gravity is actually too weak to fit into the theories of physics. Physics recognises four fundamental forces: The electromagnetic force, the weak and strong interaction and the gravity. Bold attempts to demonstrate the existence of a fifth, sometimes unknown force have so far failed[21]. 21st century theoretical physics attempts to combine general relativity with quantum physics. Quantum physics describes the behaviour of the microscopic world, i.e. the three non-gravitational forces. Here rather strange, eerie phenomena such as the tunnel effect or the infinitely fast distance effect were discovered. In particular, the latter persuaded Einstein to distance himself from quantum physics. What was happening on a small scale was too gruesome for him - although these phenomena have been confirmed experimentally as well as the predictions of the theory of relativity. The fundamental problem now is that the general theory of relativity and quantum physics in borderline cases are not compatible at all. The attempt to describe cosmic extreme phenomena such as black holes or even the big bang leads to an overlapping of the two theories, resulting in incompatible infinities,

---

[21] Some scientists suppose that to date the combination of quantum physics and the theory of relativity has not succeeded because of this unknown force.

so-called singularities. Infinities are often due to a lack of knowledge about a phenomenon. Thus the description of electromagnetic waves in the 19th century led in some cases to energies of infinite size ("oven" paradox) until the quantisation of electromagnetic radiation was discovered. It is therefore believed that general relativity and quantum physics result in a more elevated, more comprehensive theory. A so-called TOE ("Theory of Everything"), from which, hope the physicists, all the properties of the universe can be derived. From length contraction to lever law. The General Theory of Relativity, as well as the quantum theories, would be included as a limiting case for quite specific situations. Just as the Newton formulae provide very good results in everyday life or the special theory of relativity in a flat space-time with negligible gravity may be considered correct. This is where the correspondence principle comes in, an important principle of physics that says that any higher theory must somehow contain all experimentally confirmed phenomena of previous theories. For example, Newton's Law of Gravity has proven to be quite accurate for low speeds in practice. Consequently, the theory of relativity had to give the same results for sufficiently low velocities as Newton's Law of Gravity, albeit with higher accuracy. So far, however, gravity has thus far steadfastly refrained from combining with quantum physics. It seems as if gravity wants to prevent the phenomena in the black hole from being revealed and disguise and conceal its cosmic secret.

In fact, gravity is in many ways a rather strange entity that is fundamentally different from the other three fundamental forces of physics. So gravity is much weaker than all other forces. You can easily overcome the gravity of the entire earth by briefly lifting one foot or jumping around in the apartment. The other natural forces are many times stronger. So it already tantamount to a feat of strength to separate an iron chip from a magnet that may perhaps be as big as a CD. In addition, no one has yet been able to shut off gravity. In an atomic bomb explosion, however, a Faraday cage is already enough to intercept the electromagnetic pulse and protect the electrical equipment from the associated destruction. Or putting some wood on a magnet, thus shielding it from the force of attraction. But gravity works through all masses. On Mount Everest, they are attracted by the Earth and on Jupiter or Pluto the attraction of the sun is still detectable. The range of gravity is infinite according to current knowledge and penetrates all bodies. No matter what is in the way of gravity, it does not stop or shield its effect. You can lift a box off the ground and keep the box at the same height. But this is not a shielding of gravity, but only an opposing force. As soon as you let go of the box, it falls instantly back to the ground. In September 1996 the Russian scientist Jewgeny Nikolajewitch Podkletnow claimed to have observed a weight reduction in the overlying body during an experiment in Finland with fast-rotating high-temperature superconductors. A supposed indication of a successful shutoff of the gravitational field. However. this turned out to be hot air as no other scientist could reproduce the experiment. As I said, the adjustment of gravity by a (for example, magnetic or mechanical) counterforce has nothing to do with the shielding of gravity. According to the current state of science, gravity is the only one of the three basic forces that cannot be isolated. One possible explanation could be that gravity is transmitted through the structure of space-time in which all masses and energies are embedded. Therefore gravity cannot be isolated by any experiment within this structure. Another explanation is provided by the string theory, in which gravity is transmitted via ring-shaped threads that penetrate all spatial dimensions and hence also all bodies.

But that's not all by a long way. Not only can gravity not be sealed off. gravity is always attracting. In nature, we have never encountered an anti-gravitational effect. Gravity never repels, unlike, for example, electromagnetism, where the same poles repel and attract unequal poles. The gravitational force is created by masses which bend space-time and thereby exert the attraction on other bodies in the form of an acceleration. The force of attraction will always attract. If we wanted to generate antigravity we would need a mass that produces a sink in space-time, but rather an elevation. However, this is not possible with ordinary matter. Some scientists suspect that this would require matter with a negative energy density. This could create curvatures in space-time which always have a repelling effect. According to the current state of knowledge, antigravity would violate the principle of equivalence, which would require modifying the general relativity theory. Later we will go into this topic even greater depth.

Gravity is also peculiar in nature. It is actually a fictitious force, that is, its effect is due to a property of the space-time surrounding us. Each mass bends space-time, creating the appearance of attraction to all the other masses of the universe. Hitherto the range of gravity has been considered infinite, even if it becomes quadratically weaker with increasing distance. The gravity of the smallest paper clip on your desk, however, extends as far into the remotest galaxy as the gravity of the sun. A quantisation of gravity might possibly fix this. How gravity is transmitted in the end and whether it can be quantised are still unclear. Is space-time geometry already fundamental enough to make an apple fall? Or is the geometry transferred to the apple by a carrier particle? In a quantum theory of gravity, the graviton is thought to carry the gravitational force, a hitherto hypothetical massless particle, similar to the photon, which transmits the electromagnetic effect. However, no experimental evidence of the graviton has thus far been produced. Moreover, it is anyone's guess whether gravity can be at all understood like the other forces of nature and hence quantised. Moreover, and last but not least, gravity seems to be a cross-dimensional force that may permeate our space-time and reach other dimensions or even universes. This is at least one hypothesis that explains in world-formula theories like string theory or in Lisa Randall's five-dimensional universe why the gravitational force in our space-time is so weak compared to the other natural forces.

The previous incompatibility of general relativity and quantum physics could point to a far more complex nature of the universe. At least the scientists who work on popular theories of world theory such as string theory are of that opinion. However, these theories are so far-reaching that they overshadow everything that physics and mathematics have developed so far. In fact, the four dimensions of our space-time are far from sufficient, for example, to formulate string theory, a widely acclaimed theory of everything contender. The professional world basically assumes the existence of additional spatial dimensions, so the question naturally arises why we could not observe these additional dimensions in any experiment so far conducted anywhere in the world. In string theory, this circumstance is explained by coiled-up dimensions that are so small that they are undetectable in the current state of the art. From these additional dimensions can be deducted the fundamental properties of particles and natural forces. Just as gravity is the curvature of four-dimensional space-time, the effect of electromagnetism could be determined in a fifth dimension.

These dimensions currently only exist on paper. However, it would be quite conceivable that aspects of a world-formula theory in particle accelerator experiments would result as by-products, thus providing initial experimental evidence for the correctness of the theory.

The idea of a universe with more than four dimensions is by no means as absurd as common sense would seem to have us believe at first sight. Shortly after the publication of the general theory of relativity, Theodor Kaluza succeeded in uniting two fundamental natural forces. Gravity and electromagnetism. To do so, he expanded space-time by a fifth dimension and this seemed to work out well. work. Where physicists have failed for the last eighty years, Kaluza had already succeeded shortly after the end of the First World War. Albert Einstein spent the rest of his life searching for a unifying theory based on Kaluza's research. He was to fail in this like all other scientists. Because in the 1930's two weak forces were discovered with the weak and strong force, complicating the situation. The Kaluza-Klein theory extends our universe to five dimensions (four dimensions and one time dimension). This enabled him to consider the general theory of relativity and the Maxwell equations of electromagnetism as special cases. The correspondence principle, according to which a higher theory must always cover all subordinate, experimentally confirmed theories, was thus fulfilled. Oskar Klein later introduced the concept of compactification whereby extra dimensions are rolled up and are therefore invisible. The fifth dimension would therefore be a rolled up spatial dimension that is invisible to the human eye. One might think of this dimension as a purely technical dimension serving to apply electromagnetism correctly to space-time. Or one could illustrate the circumstance by means of a telephone line suspended between two poles. From a distance, the line looks like a one-dimensional line, up close like a two-dimensional surface, and only with a suitable perspective does the line really appear three-dimensional. Similar to how gravity is determined from the geometry of space-time, the laws of electromagnetism may be due to the fifth dimension. Even the Kaluza-Klein theory, however, has so far been unable to unite the general theory of relativity with quantum physics. Because it has one major snag: it cannot be quantised. The reason why the Kaluza-Klein theory was of groundbreaking importance lies in the fundamental attempt to unite the natural forces by additional dimensions by expanding the universe. In this approach, the currently most promising path to the theory of everything seems hidden. Indeed, in string theory, everything indicates that coiled dimensions, as Klein suspected in the fifth dimension, could be the key to deciphering the second-largest cosmic riddle. Kaluza and Klein may one day go down in history as pioneers of the most significant physical theory of the 21st century. Maybe someday a new Einstein with the right idea to tackle the unification problem at the root and crack this hard nut. Just as Einstein finally came to the right conclusions and interpretations through numerous preparatory works done by renowned scientists such as Lorentz, Poincaré, Riemann or Maxwell. Einstein did not simply shake the revolution of the worldview. It was the insatiable, almost pathological fascination that drove him. He internalised the findings of the previously very fruitful years and tried to combine these with the unsolved problems of the present. In years of almost convulsive effort, while working at the Patent Office in Bern, he devised a comprehensive theory to solve all the contradictions that had arisen from the Michelson-Morley experiment and the associated omission of the ether.

In summary, one could say that the theory of relativity was based on the ingenuity and ability of Einstein to combine preliminary work with his own ideas into a new overall picture and to draw the right conclusions from it. Too often, interpretation is the factor that has sealed the fate of other physicists [22].

2.2.4 The limits of the universe

For a long time physicists and philosophers were convinced of the infinity of the universe. Einstein, too, was fascinated and incorporated in his equations of General Theory of Relativity the Cosmic Constant in order to make an assumption about a stable universe. Later he described this as the biggest mistake of his life.

The question of the limits of the universe is the re-launched search for the geographical end of the world. Where seafarers once feared an abyss at the end of the oceans, today we ask ourselves what we find at the end of the universe. What happens when we cross the boundaries of space-time with a spaceship? Is there any limit to space-time, or is this notion as fantastic as the abyss beyond the Atlantic? Or is the universe simply infinite, as there are seemingly endless grains of sand on the beach and infinite stars in the night sky?

The concept of infinity in this respect is strongly influenced by the human horizon, which obscures the end of space-time as well as

---

[22] Some examples: Riemann developed the mathematical fundamentals of four-dimensional space-time. As early as the 1890's Lorentz had produced equations which postulated the shortening of bodies in motion. Maxwell recognised that his oven produced an infinite amount of energy. The really extensive theories which combined contemporary ideas and correctly interpreted approaches were developed and ultimately published by Einstein.

the amount of grains of sand on the beach, and is therefore the first to generate the idea of infinity. In fact, we can assume that our universe is not infinite, for an infinite universe leads to numerous epistemological nutmegs. So we cannot make a reliable statement about the nature of an infinite universe because we can only analyse one small, local area at a time. But in an infinite universe there are also an infinite number of these small, local realms, making it impossible to generalise our laws and beliefs to extend to the entire universe. Accordingly, a locally friendly and stable universe could be extremely hostile to life and generally chaotic. In addition, an infinite universe causes the most unlikely but possible events to happen infinitely often. So there is a place where a ready to take off Boeing 747 emerges from a cosmic accident. Since there is an infinite number of areas in such a universe, there is also an infinite number of areas where this abstruse coincidence occurs. If the universe exists for an infinite time, the night sky should be brightly lit because then there would also be infinite light that has been emitted by the stars, resulting in night turning to day. We may therefore assume that there are a finite number of stars and an equally finite number of grains of sand. There may be a lot of stars and grains of sand and we feel really impotent if we want to count them. In fact, sooner or later, even this endeavour would be crowned with success. The fact that there cannot be an infinite number of grains of sand - and in principle no mass in our universe exists in infinite numbers - can be elegantly fathomed using the following example: On the beach, we could indeed have the impression of an infinite number of grains of sand dispersed along the coast, but if we actually counted them we would eventually come to a conclusion. Of course, you could argue that no one has ever tried to count all the grains of sand, so it is quite possible that there are infinite grains of sand. You could also say that there are an infinite number of stars, and obviously there are so many when we look at the image of our home galaxy, the Milky Way. Although we have neither the technical capabilities nor a single human life would be sufficient to count all these stars, we know that there can be neither an infinite number of stars nor an infinite number of grains of sand. Yes, we can even go one step further and, in principle, determine that there can be no infinite amount of material things in our universe. Because an infinite number means that the mass associated with this is also infinitely large and would therefore generate such a high gravity that space-time literally collapses under the load and into a gigantic black hole that destroys the entire cosmos. True to the principle of René Descartes "I think, therefore I am", we can say that so far no black hole has so far sucked us up and therefore there can be neither an infinite number of stars nor an infinite number of grains of sand. After my last beach vacation, apart from an overall sunburn, I returned to my office unscathed to spend the rest of the summer typing this book rather than disappearing into a black hole. Although this derivation may seem a bit bold, we can conclude that there are in the end many stars and in the end many grains of sand - and in the end also a lot of other matter. Even if we do not have the resources to count them it would at least be possible in principle, since the result will be finite. So we can assume that we live in a finite universe. By the way, even a finite universe can seem "infinite". Assuming it is spherical, a spaceship could fly straight for an infinitely long time and yet never reach the end of the universe. An airplane can also orbit the earth as any number of times without ever reaching the bottom of the earth, as our planet is known to be round. Nevertheless, the spherical universe is not infinite because it has a definite, finite volume.

The Strange Universe: Einstein, Quantum Physics and the ToE

But is there any way for us humans to reach the limits of the universe? Can we somehow escape the universe, if we want to, provided we have the necessary equipment? Is there even a world outside our world?

A futuristic spaceship that defies all rational principles could fly at almost the speed of light, almost 300,000 kilometres per second. Of course, it is impossible to achieve such a high speed with conventional drives. Einstein's theory of relativity also prohibits the speed of light from being reached exactly or exceeded. At least not without overloading the universe. Suppose a space organisation in the distant future were to orders reckless pilot into orbit to explore the limits of space-time in his futuristic spaceship. Unfortunately, most of the General Relativity theory notes have been lost over the years. Nobody can remember them. The pilot starts the engines and prepares to explore the far reaches of our universe. So that he does not get bored and he progresses faster, he accelerates his futuristic spaceship to 99.999 percent of the speed of light. As a result, his energy reaches extreme spheres. The resulting curvature of space-time is enormous. From now on, an orange no longer rests on the film but rather a heavy stone that causes a deep depression. For the pilot in the spaceship everything goes back and forth. He happily whistles the tune of a primeval folklore song that contemporary archaeologists have unearthed somewhere. Suppose the pilot still has a trump up his sleeve. A gigantic source of energy, which he can tap at the push of a button and convert to another source, but with only very small increase in speed. He does not press the button, but the spaceship decides to play the trump in order to escape his musical efforts as fast as possible. The spaceship now flies at quasi-light speed. The pilot has half a universe under the pump, which allows him to accelerate ever faster and achieve an ever better approach to the speed of light, without ever reaching it. The spaceship is getting faster and faster and the space-time is bending more and more because of the higher energy. Suddenly everything goes dark. The whistling stops. The pilot will no longer see any light spots on the horizon. No distant stars. No galaxies flocking around voracious black holes. Even the Milky Way, which was just in front of him, has disappeared. Likewise, the distant binary system Zeta Reticuli, in which occasional extraterrestrial life is suspected. The pilot, insofar as he is still able to do so, scratches his head.

What has happened?

Let's transfer the incident to our example: As the stone gets heavier and heavier on the film, it bends increasingly. more. That means: The valley gets deeper. Once the stone is too heavy or mentally replaced by a boulder, the foil can no longer withstand the weight and tears. The stone falls to the ground. The slope of the sink is theoretically infinite. The stone has figuratively detached from and left space-time. Its weight has exceeded the critical limit that the universe can bear. If we apply this illustration to the spaceship example, we quickly realise what has happened to the ambitious pilot. He tried to outsmart nature and reach the speed of light with an almost infinite supply of energy. But the pilot apparently had no knowledge of neither the General Theory of Relativity or of the whims of nature as he was reluctant to look at the charts. Otherwise he would have guessed the result of his manoeuvre. With each new energy supply, the speed of the spaceship has increased, if only slightly, in the last decimal range. However, the energy supplied is not simply lost but has been preserved in the movement of the spaceship as kinetic energy (= kinetic energy).

The Strange Universe: Einstein, Quantum Physics and the ToE

According to the principle of energy conservation, energy cannot be lost but can be converted to other forms of energy (for example, from mechanical to electrical energy, as is the case in hydroelectric power plants). As the speed increases, the kinetic energy of the spaceship has consequently increased. Since energy and mass can be considered equivalent, the mass of the spaceship has become ever larger until it finally reaches a critical limit. A limit that may never be reached or exceeded in reality. Space-time is not a container into which you can pour and pour without it overflowing. Our universe has critical boundaries that should not be exceeded and cannot normally be crossed. The fictitious spaceship with its exorbitant (hypothetical) energy now exceeds the limit of the maximum energy that an object can have in space-time and which space-time can barely endure. As a result, space-time breaks down under the excessive load of the spaceship. The pilot, who is scratching his head does so because to a certain extent he has fallen through the filter of space-time with his spaceship. He has overcome the astronomical limits of the universe and has advanced into another existence. Into another dimension. In another universe. In hyperspace. How can something overload the limits of the universe and what happens when space-time somehow "rips"? And above all, what is hidden behind space-time, beyond the limits? Where does the stone or spaceship go when it falls out of our universe, out of space-time?

The facts appear abstract and shake the infinite concept of the universe, which is anchored at least in popular myth. The strange limits of the universe, however, are not flimsy speculations of sensationalist authors but are spawned from the great pioneer period of modern physics at the beginning of the 20th century. Max Planck, founder of quantum physics and one of the most renown physicists of all time, was one of the first to sympathise with the theories of Albert Einstein. Planck himself contributed significantly to the elaboration of the Special Theory of Relativity. However, he rejected the light quantum hypothesis of Einstein. This went too far for him, it was too revolutionary and asked too many questions of generally accepted knowledge. By publishing the Special Theory of Relativity, had already put Newton and his followers in checkmate. Planck did not want to give up Maxwell's electrodynamics, which would have resulted as a consequence of the light quantum hypothesis. It took Einstein six years to convince Planck. After a thorough study of the Theory of Relativity, Planck elaborated various equations that determined the limits of the validity of our laws of nature in their form. These so-called "Planck constants" therefore specify the limits of space-time beyond which the validity of our laws of nature cannot be guaranteed. The law of levers and length contraction are not therefore universally applicable with absolute certainty but are limited to a particular field of observation in space-time. On the other hand, this field of observation encompasses just about everything that is familiar and accessible to us from everyday life and research. Even stars and galaxies can be beautifully described. It only becomes critical for our laws of nature when we look at extreme phenomena such as black holes or the Big Bang. When we apply our laws of nature to these phenomena, the result is usually a horizontal figure eight, infinity. This is a subtle indication that our laws are incomplete and that we have reached a limit beyond which a higher theory is needed. A theory that combines the aspects of quantum physics with the theory of relativity and thus also takes into account the effects of gravity (quantum gravity). Such a theory is often referred to as the " theory of everything". In fact, many different approaches and much research is required to develop a universal formula and understand and explain our universe beyond the Planck boundaries.

In 1995, the attempt to develop a universal formula led, among other things, to the assumption that five of the world's theoretic formulae that have emerged thus far are probably just an approximation of a superordinate theory of everything, just as Newtonian mechanics from the 17th century are an approximation of the higher relativity theory. But as we have not yet discovered this theory of everything and all research efforts in this direction have resulted in extremely complicated mathematics that would occupy scientists around the globe for decades and centuries, we must realise that human knowledge of the universe and Nature ended, at least temporarily, at the Planck border. Everything that leads beyond the Planck constants can neither be proved nor refuted, and inevitably leads to speculation. This principle also applies to the inner workings of black holes as well as to Big Bang theories. But this also means that there are fundamental barriers all around us. Obstacles in Nature that block our view of any spheres. Just as we can never reach or even exceed the speed of light no matter how good our technology is. Even if we built the best microscope in the world we could only see things bigger than Planck's length. Even if we built the best stopwatch in the world we would only be able to measure time units exceeding Planck's time. Anything shorter than Planck's length or smaller than Planck's time remains locked out of this universe and inaccessible. This also means that the elementary particles do not always consist of even smaller particles, but Planck's length specifies the smallest possible length. At least until we decrypt the theory of everything and put the quantum and relativity theories under the same roof. However, the barriers are perhaps so fundamental that all ambitions to develop theory of everything are snuffed out. Clues may exist to enable us to understand the universe in all its diversity and fascination on the other side of the human event horizon. Where the pilot landed with his spaceship. Nevertheless, there are already continuing world theories which could also be very informative in terms of the existence and the cause of the Big Bang (more about that later). However, many things are still very vague because experimental evidence in a sphere that exceeds our space-time by many dimensions is correspondingly cumbersome complicated and energy-intensive. The cause of an absurdity, which one or other reader might have noticed after the first few lines of this sub chapter, is also necessarily extremely energy-intensive. Recent NASA measurements estimate the age of our universe[23] at 13.7 billion years and the diameter of the same universe at around 93 billion light years[24]. According to the theory of relativity, no matter and no information can propagate faster than light[25]. Consequently, the cosmic background radiation (and depending on the interpretation of the universe as such) is not. As a result, the universe may only have a maximum diameter of 27.4 billion light years. What is the explanation of an almost quadrupling of astronomical observations in 2008? The current standard model of the Big Bang suggests that there was a supraluminal expansion at

---

[23] To avoid confusion with other possible universes I occasionally refer to the known universe as "our" universe, but without wishing to derive any claim of ownership from this.

[24] One light year is equal to the distance which light covers in one year (at the speed of light).

[25] Particles which move at supraluminal speed and are wholly compatible with the general theory of relativity. However, such particles always move faster than light, can never be decelerated to the speed of light and do not even drop below it.

the moment of birth of the cosmos, a gigantic cosmic inflation in which space-time has developed to some extent. At least this phenomenon would not be incompatible with our physical view before Planck's time (10^{-44} seconds after the Big Bang) because we cannot make reliable statements about events outside the Planck limits based on our laws of Nature. A much faster supraluminal expansion outside our horizon might very well be possible. In fact, an excessively fast expansion would have abstruse consequences if an attempt were made to explain it on the basis of the general theory of relativity, that is, by the modern physics we know today. Namely, any particle exceeding the speed of light would move backwards in time. At the time of the Big Bang in a physical past, which has not happened yet (based on the Big Bang theory, the expansion of space-time began with the Big Bang, so "our" time did not even exist before that time).

If the astronomical observations are correct and the universe actually has a diameter of nearly 100 billion light-years, the propounded inflation theory would be a possible explanation. On the other hand one could also invoke possible exotic forms of the universe and resulting incorrect estimates of age or diameter as explanatory approaches. Another possible alternative would be the realisation that we have not yet sufficiently understood Nature in order in turn to understand such comprehensive events as the moment of birth of the universe or to determine scales such as the dimensions of the universe. The boundaries of the universe may remain an interesting, entangled field of research for quite some time.

2.2.5 Escape from the universe

Anything that exceeds the limits of scientific knowledge has always had a special attraction for humans, as has the escape from our universe. From the domestic space-time. From our world. As impossible as it seemed to our ancestors to fly like birds through the air, today it seems impossible to escape from the universe.

Let us recall the words of the German physics professor Johann Christian Poggendorff in 1860: "It is impossible to transmit speech. The 'telephone' is as mythical as the unicorn" [26]. Of course, we have not discovered a unicorn, but the phone has revolutionised communication. Through the phone and its evolution the world has come together and it as natural today as the sun rising daily.

On the basis of modern physics it would not be possible to establish a stable theory about events outside of our universe. For in doing so, we inevitably enter areas in which quantum gravity, a union of quantum and relativity theory, attains significance. Nevertheless, a consistent continuation of the guiding principles of modern physics enables some astounding conclusions and theories to be propounded about the world beyond the bounds of our universe.

An escape from our universe is basically conceivable in three different ways:

1. Exceeding the Planck constants

2. Exceeding the spatial extent

---

[26] Translated: It is impossible to transmit speech electrically. The telephone is as mythical as the unicorn".

3. Cosmic escape route

The first case corresponds to the path the ambitious pilot took with his spaceship in the previous sub-chapter. Exceeding the Planck constant leads to an overload of the space-time structure, so that the object in question is somehow extracted from the universe. This can be illustrated with the heavy stone or boulder which is placed on the taut aluminium foil. The foil is overloaded, rips, the stone falls through the foil. In practice, however, taking into account 21st century technology, this escape attempt is hardly feasible. To go beyond the Planck limits would require reaching extreme states, which can often only be brought about by other extreme influences. The pilot, for example, needs an exorbitant, almost inexhaustible source of energy to drop his spacecraft through space-time. The necessary prerequisites may be illustrated, for example, by the concept that instead of the stone, an entire peak chain, let us say the Himalayas, would have to be put on the foil in order to tear it. The space occupied would probably be quite small, not to mention the other circumstances that rather complicate the experiment. The second way is not very promising. The edge of the universe would never be reached, let alone "overtaken" at the speed of light, or even in billions of years. Moreover, it is more than questionable whether there is any kind of "external territorial border" of the universe. To enter the US, we can climb over a border fence in Mexico. However, to escape from a bullet we can run our entire life in one direction and never arrive at the target (the surface of the bullet). According to our conventional understanding, it should be possible, at least theoretically, to overtake the expansion of the universe or the outer boundary by "simply" moving in the same direction faster than the speed of propagation of the universe. Sooner or later you would have reached the outer edge in this way. If Karl left ten years ago, you would have to catch up with him sooner or later with a fast sports car, provided his path did not lead him over hill and dale, otherwise the only thing that overtakes is our front wheel. Einstein has already provided adequate evidence that our everyday experience in astronomical dimensions often fails us. According to our intuition and the doctrine of Newton, it would in the end easily be possible, at least on paper, to catch up with the light by simply moving faster and faster. The constancy of the speed of light, a postulate of the special theory of relativity, has shown us that we should not engage in a race unless we know a pretty good short-cut.

Our Universe is most likely a complicated, multi-dimensional construct of which we have been able to measure only three spatial dimensions so far. The time dimension as the fourth dimension, strictly speaking, cannot be directly observed. Although the universe is objectively quite finite, we three-dimensionally based beings cannot cope. even if we could fly through space-time with light speed and billions of years in our luggage. Let's illustrate the problem with a small example: an ant walking on the surface of a sphere or a circle-closed garden hose (similar to a doughnut). Te ant can move as it pleases. It will never reach an end. From the perspective of an outside observer, of course, the sphere is finite. It has a certain volume, mass and surface that we can easily grasp with our three-dimensionally trained eye. From the perspective of the ant, however, the spherical world never ends, that is, it will never arrive at a visible end. At best, it squirms in a circle. Similarly, though more complicated, it could be related to our universe. There we are the inconspicuous little ant that cannot escape its world. We have only a very limited view, knowing three spatial dimensions, and perceiving the universe from a local, inner perspective. The universe may be a spatial prison for us, from which we cannot easily escape. Not because there is no way out, but because it is possibly a multidimensional and complex entity that we cannot perceive in its entirety. We are on the inside of the structure, like the ant in the garden hose or on the spherical surface. If, for some obscure reason, the expansion of the universe were to one day stand still, we might one day reach the virtual end of the universe, the outer edge. However, there might not be a visible end like the sphere. We would move like the ant in a circle and sooner or arrive later at the starting point if we just kept on flying. Quite apart from the fact that we will probably never reach the end of the universe, this variant is enough to be to leave space-time. This leaves the last alternative, the cosmic escape route. By this we mean astronomically extreme phenomena which lie somewhere between "belonging to our universe" and "belonging elsewhere". These include, for example, black holes or wormholes. Returning the ant, in to mind, it can only escape theoretically from its spatially geometric prison: it can create a passage into the interior or exterior of the ball or the garden hose. This would allow it to leave its limited scope, the surface of the sphere (or possibly the interior, depending on where we "place" the ant), and move into another, new sphere. The situation is similar to the "cosmic escape routes", as will be shown in the sub-chapter "Black Holes". But what happens if one of the three ways actually leads to the intended destination? What happens when someone leaves the universe? What is behind or round our space-time? Gaping emptiness? A cosmic nothing?

A popular representation leads to a higher universe, so-called "hyperspace", space-time. Hyperspace is expanded by at least one spatial dimension [27], which would probably create a rather obscure impression on human visitors. Your mind would not be able to grasp the fourth space dimension (time is not a spatial dimension). A four-dimensional ball might seem like a disc and a glider of an extraterrestrial civilisation like a cut-off, faulty polygonal representation in a computer game. At first the occupants of the "emigrated" spaceship are likely to feel rather confused and disoriented in this higher universe. Not only is visual perception subject to fundamentally altered principles, but the physical properties of the universe are likely to play out on a different level. For example, the speed of light in five-dimensional space (four extended space dimensions and one time dimension) could be significantly higher due to the extra degree of freedom than in our universe. Thus, the Planck constants were calculated to other limits, since four degrees of freedom are likely to create completely different conditions. Of course, the fourth degree of freedom exceeds our spatial imagination. Similarly, it cannot be imagined how hypothetical life forms of hyperspace would be. Perhaps these beings would have far more powerful brains, since understanding four spatial dimensions is much more demanding. The brain of homo sapiens was already increasing noticeably as the collectors and hunters learned to throw. How superior would civilised beings of hyperspace be to us? From the point of view of these life forms, we would probably be trapped in a world that we could never leave without extreme actions or external help, let alone recognise in their higher-dimensional condition. As opposed to the ant, which

---

[27] Five-dimensional hyperspace must not be confused with five-dimensional space-time, which is postulated, for example, in the Kaluza-Small-Theory. The fifth dimension of hyperspace is an accessible, macroscopically extensive space dimension like the three space dimensions we know about (and not a microscopically coiled dimension).

comes only from or into the ball, as soon as an oversized tunnel or a hole appears as an escape route. On paper there are various ways of opening a gate into this hyperspace. If you succeed in crossing the natural limits of our space-time, an escape from the universe is conceivable. Wormholes, black holes, absolute zero or acceleration to the speed of light - at least the conceivable emergency exits are supposedly numerous. However, in many cases the necessary energy exceeds the theoretical range. As it would not be possible with conventional engines ever to accelerate to approximate speed of light, dropping through the sieve of space-time by established methods is almost impossible to accomplish. To be able to leave the universe or space-time, a fundamental re-think is essential. An escape into hyperspace is certainly not feasible using a battering ram and crowbar. But if we develop an understanding that allows us to have a more objective view of things, we could reach for the key. A more objective understanding of things goes hand in hand with a higher-level theory that explains and understands how extreme cosmic phenomena actually occur. Only through achieving this does it appear conceivable to open the doors to another universe. The ant remains a hopeless prisoner of its space until it has a more objective perspective of an outside observer and is able to re-evaluate its world based on these findings. The world suddenly no longer appears so infinite to it and it could rid itself of the naive idea that continuing perpetually in the same direction inevitably leads to the desired destination— that long-awaited other world. Perhaps the general theory of relativity already contains a key to other universes. After all, not all the possible solutions that can be calculated from formalism are known, and the solutions we already know make possible phenomena such as the Einstein-Rosen bridge. A tunnel through space and time.

2.2.6 The Einstein-Rosen Bridge

In 1935 Einstein and Nathan Rosen discovered a special solution to the field equations of the General Theory of Relativity: the Einstein-Rosen bridge – a link between two points in the universe produced by a gravitational anomaly. A tunnel through space-time. A bridge to the most remote areas of the universe. A potential connection in parallel universes, hyperspace or distant points of our space-time. This tunnel is commonly called a wormhole, on the analogy of a worm that eats through the apple and thereby shortens the distance to the other end of the fruit.

The ambitious pilot of the spaceship, which has fallen from space-time, could therefore use a wormhole to get from the Earth in no time to another galaxy or a parallel universe. A wormhole is not the offspring of fantastic scriptwriters or thriving fantasy, but is a cosmic entity that derives from the General Theory of Relativity. Wormholes are as real as the gravity that keeps us on the ground every day. A wormhole is a tunnel-like link between two universes or distant space-time domains. I emphasise that it is a tunnel in space-time, and not just in space. Since the three dimensions of space are inseparably interwoven with the dimension of time, every movement in space is also associated with a movement in time. Depending on the geometry on which our universe is based, or what our universe looks like from the outside, a wormhole may also signify a tunnel through time.

So far we have only an approximate idea of the structure of the universe. Therefore, we do not know what effect an Einstein-Rosen bridge has on the passage of time and whether this makes it possible to move directly in time. It is conceivable to create wormholes on Earth. As a result, spaceships could advance within the shortest time into distant and previously inaccessible areas of the universe. Or even in other universes, like hyperspace. Wormholes could be cosmic escape routes. However, a connection between two universes or space-time areas through an Einstein-Rosen bridge does not arise out of nothing.

To ensure that a stable wormhole can be constructed exotic matter is necessary according to the theoretical physics.

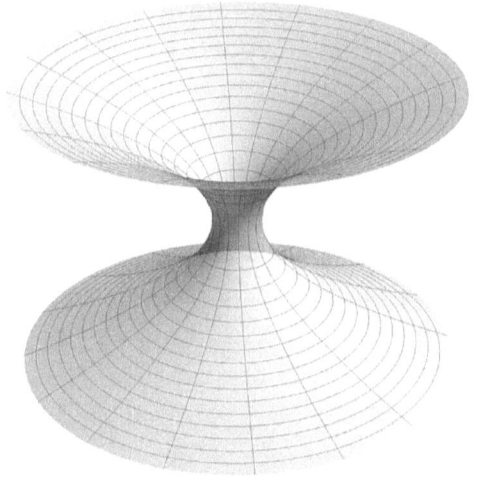

Figure 4   A wormhole in space-time

But what is this exotic matter? Exotic matter does not consist of ordinary atoms and elementary particles. Consequently, in principle, the dark matter and dark energy are considered exotic due to their unknown properties. Dark matter and dark energy are cited as explanations for cosmic phenomena, which cannot be satisfactorily explained even by general relativity. The universe is said to 95 per cent filled with dark energy and dark matter. But the building blocks for wormholes must lie elsewhere, otherwise our universe would be riddled with Einstein-Rosen bridges. In this case, the astronauts of the Apollo missions might not have landed on the moon, but in a distant galaxy in a possibly distant time. Exotic matter could also be matter with negative energy density, as it would be needed to generate antigravity. However, the existence of such matter has not been proven experimentally.

Whether the Einstein-Rosen bridges will allow travel to other solar systems and galaxies in a few decades or centuries is still unclear today. Some researchers estimate the need for exotic matter equal to the mass of the planet Jupiter to create a stable wormhole one metre in diameter. Jupiter is about 318 times heavier than Earth or about 2.5 times heavier than the rest of the Solar System. Other estimates are more modest and even suggest small amounts of exotic matter as quite sufficient to open the door to intergalactic voyages. But no one can say where an Einstein-Rosen bridge will lead in space-time and whether a return is ever possible. The only thing that is clear is that a wormhole, purely in principle, allows intergalactic travel and thus represents one of the only known options to reach other galaxies in a very short time. A wormhole could transport the pilot with his spaceship in a few moments from the moon directly to the centre of the Milky Way - which is not necessarily recommended, as the presence of a black hole is suspected in that space. In just a few seconds, the pilot could cover a distance of about 27,000 light years through this wormhole. A gross violation of the laws of nature? Not at all.

The pilot does not exceed the speed of light as the maximum speed allowed in our space-time. A wormhole represents a kind of short-cut that connects two possibly very distant areas of space-time. The worm does not have to crawl over the surface of the apple to get to the other side. It can simply eat itself and thus shorten the distance. The distance between the Moon and the centre of the Milky Way is correspondingly shortened to the "length" of the wormhole. That such short-cuts are possible could also be due to a higher-dimensional structure of space-time. In any case, the pilot does not move at supraluminal speed but simply shortens the route. Incidentally, the Einstein-Rosen bridge has recently been firewalled against the string theory, one of the most promising candidates for the theory of everything. Here it has been shown that cracks in space-time are possible. However, these cracks do not lead to the catastrophe in which matter escapes from space-time, but space-time heals the rift with a new connection. This may create a networked and "wild" structured universe - much like the interconnections in our brain. That would mean that our local perception of the consistent, flat universe is wrong. Rather it could be traversed by accidental Einstein-Rosen bridges connecting space-time areas. This in turn raises some important questions about the credibility of astronomical observations and calculations. For example, how can we roughly determine the age of the universe or the distance of stars and galaxies when the light from these celestial planets can be tunnelled through a wormhole? If the universe is peppered with Einstein-Rosen bridges, our perception of the cosmos may have been completely distorted. And with it the image that we have of the universe.

But we have not discovered a wormhole yet. Perhaps we should focus on finding clues, as wormholes are unlikely to be directly visible, unlike popular work-ups in science-fiction movies. It is not a celestial body like a star emitting light, but a combination of different space-time areas that shorten distances, much like a motorway tunnel through a peak, which spares you the detour via the winding pass. In addition to physics, a philosophical argument advocates the existence of wormholes. Why should a universe consist of billions of stars and galaxies that can never be reached by a civilisation? If the existence of man has meaning, and is not based on an extreme coincidence, then would not the existence of Nature also make sense? In principle, wormholes enable us, even if we do not yet have the technical knowledge to do so, to penetrate these areas of the universe. With conventional fuel or nuclear drives, this project would be doomed to failure due to unimaginable distances from the outset. The nearest star system "Alpha Centauri" is about 4.3 light years from Earth. The Milky Way, "our" galaxy, has a diameter of an estimated 100,000 light-years. It is about 2.4 million light years to the nearest galaxy "Andromeda". The light of the stars from the Andromeda galaxy is travelling 2.4 million years before it arrives on Earth[28]. In this respect, we look back 2.4 million years into the past when we point our telescope at this galaxy. As we have not "built" or travelled through a wormhole, we do not of course know if humans could use these cosmic short-cuts. There are theories stating that Einstein-Rosen bridges exist only in microscopic form and are thus inaccessible to humans. It remains to be seen which theories correspond to reality.

If wormholes in the Universe were to occur more frequently than expected, this might give rise to an attempt to solve some of the cosmic riddles that occupy us today. Certain hitherto unexplained phenomena, such as dark energy, could at best be attributed to observation errors caused by hitherto unknown short-cuts in space-time which are made by light and matter. It is not clear what it

---

[28]   Moreover, it would have been tunnelled by an Einstein-Rosen bridge ....

would mean for our world view and astronomy, should it one day turn out that our universe is riddled with Einstein Rose bridges. While we do not know if Nature has been created so that we humans can discover it, we can be pretty sure that wormholes and space-time anomalies of all kinds are the only foreseeable ways to overcome cosmic distances within a reasonable time. And to delve deeper into the wonderful world of the universe, Nature.

2.2.7 Time travel and the fourth dimension

Since we renounced caves, clubs and axes, society has been fascinated by time travel. Many of us have dreamt of a time machine that meets normal demands in terms of boosting our luck at the lottery or to give a helping hand when disappointed in love. Since the earliest civilisations, and perhaps far beyond, humanity has been concerned with the essence of time. Would it ever be possible to penetrate the future or the past, to manipulate the history of humanity or even to provoke a paradox of time with unforeseeable consequences?

At least since Einstein it has been clear that time is not an arbitrary spawn of humanity merely to bring order and structure in everyday life. Time is an invention and peculiarity of Nature. Time is the fourth dimension. In this chapter, we will discuss the possibilities afforded us by the General Theory of Relativity in its formalism in order to move through time. Time travel opportunities that do not spring from any science-fiction novels but are found in modern physics and therefore in Nature. The theory of relativity destroys Newton's linear views. Time is no longer an absolute that passes equally for everyone. One second on the sun takes longer than on earth. Time is relatively and significantly influenced by acceleration, speed and gravitational fields. This applies to elementary particles as well as spaceships or planets. A person who spends his entire life riding a motorcycle on California's Highway 1 will hardly notice the life-prolonging effect. Indeed, time actually passes more slowly for him because of his speed and the gravitational field, which is stronger on the Pacific than in the high peaks. However, the speed is so low compared to the speed of light that it causes a time delay of a few fractions of a second - calculated over a lifetime. The same applies to the gravitational time dilation. However, when we move from the California coast to the experiments using particle accelerators, the predictions of the general theory of relativity are confirmed in an impressive manner because here particles can be brought to speeds at which time dilation is very visible. The life expectancy of an elementary particle called a "muon", for example, increases tenfold to near-speed of light during acceleration. This effect is due to the enormous speed and the resulting relativistic phenomena. However, time dilation cannot only be observed in particle accelerators but also in cosmic radiation, which constantly strikes the upper layers of the earth's atmosphere at a height of about ten kilometres. This produces muons that fly to Earth at approximately the speed of light and can be detected there. Immobile muons disintegrate within about two microseconds and, according to the classical conception, only manage to reach a distance of about 600 metres. They should not reach the earth's surface at all. The fact that they can still be detected on Earth is due to time dilation and length contraction, i.e. the relativistic effects that we know from the theory of relativity. Precise measurements prove that very fast muons do not decay until about 13 microseconds, about six times slower. Impressive proof of the correctness of the theory of relativity. If it is possible to transport muons in particle accelerators to some extent in their own future, would it not then also be possible to make time travel accessible to people? What happens when a spaceship flies through the universe at almost the speed of light? In this case, practically no time should pass for the astronauts and they should not therefore age. Is a spaceship travelling at the speed of light an infinity serum? The assumption is quite correct. If a spaceship moves at the speed of light, time stands still for the spaceship. However, we know from the special theory of relativity that no mass can reach the speed of light since an infinite amount of energy would be required. The occupants would actually cease to age. All clocks and processes on board stop. But it would be presumptuous to assume that the astronauts would expect an infinite life in the everyday sense in a spaceship travelling at the speed of light. Rather this endeavour leads the astronauts into the future, depending, at best, on how long the spaceship retains its speed. However, there is a minor snag here. Freedom of action is restricted to the same extent as the passage of time. When an astronaut wants to scratch his head, it takes an infinite amount of time to even move his finger. In fact, he could not even imagine the idea of scratching his head, since the transmission of brainwaves would also take forever. Time standing still also means that events stand still. The astronauts do not age, but their freedom of action is restricted to zero. When time stands still, such as on the event horizon of a black hole, each activity takes an infinite amount of time to complete. Of course, this does not only apply to people and watches but also to all devices and every process within the spaceship. Once the spaceship flies at the speed of light, it could never be stopped again of its own accord. Neither by a computer nor by manual control, as each of these actions takes an infinite amount of time and would never be done.

Well OK. Since no matter can be accelerated to the speed of light anyway, we will focus on the question of what practical consequences it has for our astronauts when we fly the spacecraft at close to the speed of light. Almost, but not quite at the speed of light. In this case, the time on board is relatively slow but does not stop. The astronauts are ageing, but much slower than for us, who are waiting on Earth for the things that are to come.

In order to find out if a spaceship is a suitable time machine for transporting people into the future the space agency is planning a mission to answer that question. Astronaut Peter is always efficient when training at the training camp and reads exactly one non-fiction book on biology, chemistry, physics or sorcerer apprentices every day. If he travels to distant stars, he also wants to know something about the condition of Nature and thus better understand his discoveries. Or something like that. For one reason or another he has in any case read exactly one book each day, processing exactly 200 works of more or less uplifting literature in his 200 days of training. Now Peter is sitting in the spaceship and flying at almost the speed of light through space in search of distant galaxies, stars and planets. Before launching the spaceship for his mission, Peter threw some test equipment and instruments out of the hold and replaced them with a load of books to keep him from getting bored on the long journey. Mission Control threw out his books and replaced them with one e-book. In times when spaceships travel to other stars you should not waste valuable space with heavy books. The spaceship takes off and flies away at high speed, so that time in the spaceship passes ten times more slowly than on Earth. When Peter reads on his watch that he has been travelling for four years, forty years have passed on Mother Earth. How many books can Peter read in the spaceship during the first 200 days of his journey?

In order to answer this question, we need to determine the perspective from which we calculate the days. If 200 days have passed on Earth, he will not have read 200 books, but only 20 books. If 200 days have passed from Peter's point of view, he will have read 200 books, but on Earth it will have been 2000 days or more than five years. Although time passes much slower for Peter in the spaceship he cannot read more books in the same time than at the training centre. This peculiar circumstance has to do with the fact that all processes that occur in the spaceship are affected by time dilation. Everything that happens on the spaceship is recorded in slow-motion[29]. On the one hand, Peter is ageing more slowly, increasing his life expectancy from eighty to eight hundred years. On the other hand, it takes ten days instead of just one day to read a single book. He lives longer from the point of view of people on Earth, but cannot do more things within this time than an ordinary person on Earth. But the spaceship is definitely a time machine into the distant future.

One fine sidereal day Peter wants to renounce his monotonous loneliness in the spaceship and decides to return to Earth to spend his old age in the bosom of his family and friends. After all, he has grown old by now. He turns around and flies full throttle back to his homeland. On arriving back on Earth, Peter is irritated by what has changed. The oceans are populated, there are flying cities in orbit, the continents have shifted, the coasts are not recognisable and on Greenland there is a Caribbean flair. As he lands his spaceship, an army of archaeologists and historians arrives to marvel at this colossal work of ancient engineering. A tourist

---

[29] Unless otherwise indicated, the statements must be understood from the perspective of a (broadly) motionless observer, for example from the perspective of the space agency on Earth.

suspects an elaborate construction for the carnival. Another considers Peter to be a Goth who is indulging in the good old days, which are known only from the melancholy echoes of history. What happened? According to his logbook, Peter left for his mission in 2050 and then flew through the cosmos for fifty years. He visited different stars and planets. However, as time passes more slowly in the fast spaceship, five hundred years have passed on Earth, while Peter's calendar indicates the year 2100. In fact, it is 2550 on Earth. While Peter has aged fifty years, he has travelled 450 years into the future. To some extent, he used time dilation as a time machine to push forward into a future he would have found impossible to experience on Earth.

Peter's story is a good example that illustrates the importance of the theory of relativity. For people on Earth, Peter lives in slow motion in his spaceship. For Peter, the people on Earth live in fast motion, so they seem to move much faster. This is at least the effect of the gravitational time dilation that occurs during acceleration. Peter needs ten times longer for the exact same action as his equals on earth. Conversely, Peter would feel that humanity is moving ten times faster than usual.[30]

So how you, dear author, possibly claim in the chapter on the theory of special relativity that the perception of the effect of time always depends on the observer whilst at the same time claiming that it can be used to build a time machine? From the point of view of humans on Earth humans it is astronaut Peter who lives in slow

---

[30] Here a factor of 10 must be understood as a sample value. Depending on the exact circumstances, this value may turn out to be lower or even considerably higher. The closer the spaceship comes to the speed of light the greater will be the time dilation and the further Peter is able, in principle, to advance into the future of the Earth within his life.

motion, while Peter sees the quite the reverse situation: the people on Earth move in slow motion. How is it then possible for Peter to return to Earth and age more slowly than people on Earth? Are both reference systems equal? From your point of view, should not people on Earth have aged more slowly?

This apparent contradiction is known as the "twin paradox". Hans and Peter are twins. Hans stays on the ground while Peter flies in a very fast spaceship to a star and back to his home planet. When he arrives back home Peter realises that Hans has aged 50 years, whereas Peter has aged only 10 years. From the point of view of Hans, would not the situation have reversed, wherer would have aged 50 years while for him only 10 years would have passed?

Basically, this objection is of course justified. The relativity principle in the special theory of relativity states that all reference systems are equal. Accordingly, Hans and Peter are of course right in their observation that the other is moving in slow motion. But how can it then be explained that Peter has aged more slowly in the spaceship than Hans?

In fact, the effects of time are only interchangeable as long as the movement is linear and uniform. Because only then are the reference systems equal and cannot be distinguished from stationary systems. The details of the solution to this riddle are hidden. The movement must be linear. But as soon as Peter reaches the star and returns to Earth, he has to turn his spaceship round. As a result, the movement is no longer linear, so the effects of time are no longer interchangeable. The time dilation is no longer dependent on the point of view, but an effective impact develops on which both observers agree: Peter has aged fifty years, but Hans only ten years. The gravitational time dilation, i.e. the time expansion that occurs in accelerations and in gravitational fields (as explained in the chapter on the general theory of relativity) always has an effective impact on which the observers agree.

The twin paradox is therefore merely an apparent contradiction. The knowledge gained about the relativistic effects of time do not currently allow constructable time machines. Yet. In the distant future, when it should be possible to accelerate people to near the speed of light, the phenomenon of time dilation could be used as a kind of passive time travel. You could fly a spaceship with autopilot enabled to a distant star. The ageing process of human occupants would be extremely decelerated by time dilation. After a few thousand Earth years, the destination would have been reached and the artificial ageing machine could be brought to a standstill. Basically, of course, the autopilot and the on-board technology are affected by time dilation. This can have fatal consequences. Due to time dilation, the time in the spaceship passes so slowly that an approaching obstacle would only be recognised if it had already rammed it. The journey must therefore be meticulously planned and coordinated before departure. The autopilot would have to be programmed to initiate the various manoeuvres at an early stage, taking into account full time dilation - so that at least the known obstacles and planets are avoided. The astronauts would have travelled some thousands of Earth years upon reaching the destination and would have barely aged, while whole generations have passed on Earth. The space agency that once sent the spacecraft on its long journey might not exist any more. But the astronauts would still be young guys who have passed several thousand years of human history in flight.

In addition to this rather passive and elaborate form of time travel, which knows only one direction and therefore cannot be reversed, it could theoretically be possible to move freely in time. Both in the future and in the past. At least mathematical solutions of the general theory of relativity point to this. Since time is a physical dimension and forms an inseparable structure with space, it could also be accessible or manipulable in some way. The main problem with this is the question of how to interact with the time dimension. A movement in time also always means a movement in space, since space and time in space-time are inextricably linked. There are some basic questions must be answered, at least superficially. What is the time dimension anyway? How can you influence, pass through and manipulate it? How do we access the properties and characteristics of this extraneous dimension? We are evidently facing a similar dilemma as the ant in the sphere. If we want to move in a three-dimensional space from A to B, we can sometimes do this on foot, where we walk in the desired direction, provided there is no peak in our way. But how can we move in a dimension that we can neither see, feel, taste, hear, touch nor even grasp with the senses? A dimension that knows no spatial extent and is based on a completely different level of perception. Quite generally. What impact and unpredictable consequences does a journey into the future or past have on history? What is reality? What is the present? And how can we escape the fourth dimension again?

The division into present, past and future is nothing but an observer-dependent assessment of time. It's all a question of perspective. The ancient Egyptians certainly defend their presence as the true reality to the best of their knowledge and conscience, even if a dubious visitor from the distant future were to arrive in a time tunnel. Similarly, Attila king of the Huns was confident that he was leading his campaigns in the here and now and not in any historically worked-up past. Likewise, the men behind the time traveller feel at home in their present and regard the legendary conquests or the construction of the pyramids as the past. The key question here is whether the present moment depends on the observer or whether there is an absolute present as the only instant of time when history is written. Is it possible to watch the Egyptians build the pyramids, even though that time is long gone? But what happens when a case that is actually impossible occurs? If a consequence prevents its own cause? An event that is incompatible with temporal logic and the principle of causality? How does Nature react when a time traveller brings our technologies to a civilisation that lived 5,000 years before Christ? What happens if he kills someone or someone from his own family tree and is never born? Does history know a mechanism to self-correct this? Otherwise, would not a journey back in time suffice to throw the entire historical development off balance? On the other hand, does the chaos theory say that the flapping of the wings of a single butterfly can trigger a tornado? Does a logically incompatible modification of time lead to the formation of a parallel universe, a kind of copy of our existence in which history continues, taking into account the fatal interventions in time? Is not the present the absolute reference system par excellence, the only universal moment on the timeline in which something can happen that is not already past or future? Is the course of time relative to the observer, but not the present moment in itself? Can there be only one present, the moment you read these lines, or is it conceivable that the present is just a view of a particular reference system?

Causality is one of the fundamental principles of general relativity and science in general. It says that the cause always occurs before the effect, since the information can propagate only at the speed of light. The effect always has to be based on the cause. If, on the other hand, it were possible to fire a missile at supraluminal speed, the target would be in ruins before the rocket was launched. At least from our experience, we know that an effect cannot overtake its own cause. Nature prevents such causality problems by not allowing information to be faster than light. Whether this is a fundamental principle, we have not been able to say with certainty until today. For, as quantum physics and the theory of relativity have taught us, even intuitive and self-evident assumptions may turn out to be wrong when examined more closely. So let us dare to conduct the thought experiment and ask ourselves what happens when someone gets into the past? And is it even possible to live in a time other than our present?

Actually, we know very little about the nature of time. Perhaps past, future and present are fictional, man-made states to bring some structure and order into our experiences and expectations. From the past, we remember what we experienced, we set expectations and plans for the future. However, scientific time as a fourth dimension may not know the difference between what has been and what will be, as these definitions may depend on perspective. As the three dimensions of space and their degrees of freedom always exist simultaneously - there is currently no indication that the number of spatial dimensions can be locally reduced in any way - past, present and future may also always be connected. Past, present and future are therefore not on a fixed timeline, leading only in one direction, but are constantly superimposed and therefore take place simultaneously. Just as they can go left or right, up or down, forward or back at any time since spatial expressions always exist everywhere in space-time. Considering this contradictory theory at first glance, however, some apparent paradoxes can be eliminated, assuming we start from a world in which time travel back in time is possible in principle. An example of this is the well-known father-son paradox. If Hans travels to the past to kill his father Max before he has conceived him, Hans should never have existed. For with this Hans has destroyed his cause with an effect that, according to the principles of causality, should not exist. The offspring does not come from the stork after all. The murder gives rise to a paradox of time that violates causality and disfigures the logic of the rest of history. If the universe possesses a kind of "inner consciousness" that oversees the logic of things, Nature could possibly sustain itself. How this would look in the concrete case is, however, the burning question. On the one hand, the correction could be that Hans never existed and Max was shot dead by someone who never existed. A murder without perpetrators. For Einstein, a cruel murder of causality. On the other hand, it is conceivable that a guardian of the time would solve the problem by not letting it arise. Perhaps the cosmos has an integrated protection mechanism that prohibits interventions in time. The past would be past and immutable for all time.

It would be possible to travel into the future through time dilation and other anomalies, but not to reverse that journey. Connected with this is an absolute present, so that the event of that moment in the entire universe takes place at exactly the same time, according to which theory no one may and may not be in the past or in the future (since the future is the past of the present).

An elegant solution to the contradictory problem is the unification of past, present and future. Assuming that these conditions actually all occur simultaneously, history would basically know from the year zero what Hans will do in 2000. There obscure complications might not even arise. For example, the murder of Max would be anchored in history. Perhaps the people in his immediate environment would even consciously perceive that something was not going right or left, allowing someone from the future to intervene in the event. Free will would not be restricted anyway. The present and the future take place at the same time. If Hans in the present wishes to kill his father, history would have planned that from the beginning, but not so much by history as by Hans himself. Perhaps the relativity in the time dimension also exists, but only conditionally. It is conceivable that everyone has an absolute anchor point or history itself has an absolute anchor point. Accordingly, interventions in the past would be possible. The past, however, would be revived only as long as a time traveller is in it, and would have no lasting impact on other people's past. For the time traveller, the past would somehow become the present (the present is a question of definition: for a person, the present, whatever its place in space-time, is always the moment that is happening, regardless of whether the person is historically fixed in the past or in the past future). Changes in the past would have a spontaneous impact (for example, Max could fight Hans and hurt Hans). Any changes, however, would have only local influence and no far-reaching consequences for history as experienced and lived by the rest of humanity who did not travel in time.

The "multi-world theory" from quantum physics explains the mysterious phenomena of time with the realisation of every possible event in a parallel universe, so that logical contradictions are excluded, even in the case of seemingly causality-injurious events. This means that there are an infinite number of copies of our universe. Whenever decisions have to be made, a new universe is created for each decision variant in which the decision is implemented. Thus it would be possible for Hans and Max to live together peacefully in one universe, whereas in another form of the universe Hans was not even begotten. Likewise, World War II could have been prevented in one universe and the Cuban missile crisis in another universe could have escalated into a global nuclear war. The distance between these decisions could be defined by Planck time. To a certain extent this time span marks the shortest possible time span in the universe. Since Nature follows a minimal principle in its laws, it would also be possible for the universe to split each time a potential for contradiction arises - for example, when Hans travels to the past. Also in this case paradoxes would be prevented since in each universe reality would be implemented in such a way that history emerges as a consistent sequence of events. Since every piece of information reaches our senses at most at the speed of light (and since, moreover, the human response time is around 40 milliseconds), in principle we always only perceive the past. In order to answer the question of time travel into the past, we first need to understand more fully what time really is. To travel to the future, we could take advantage of the effects of time dilation. Travelling into the future would be possible, at least in principle, even if we do not yet have the technologies to carry a person far into it. Whether it is even possible to travel or influence the past is still unclear. However, there are indications that Tachyons exist, special particles that always fly at supraluminal speed and therefore move backwards in time.

2.2.8 Tachyons and the past

The first time journeys into the future are already past. Far away from the media presence, small particles started to fly into the future in particle accelerators. Not only muons and other elementary particles but also humans have already travelled to the future. Without this being reported in the press. Sergei Konstantinovich Krikaljow is the person who travelled the furthest into the future. The Russian citizen was born on August 27, 1958 in Leningrad. In April 1989, the crumbling USSR proclaimed him a "hero of the Soviet Union". Three years later, Russian President Boris Yeltsin even promoted him to "hero of the Russian Federation". Krikaljow participated in six space flights and spent over 800 days in orbit. Due to the high speed of the space station, he travelled about one fiftieth of a second into the future compared to the stationary earth population.

Humanity knows about the principal methods of travelling into the future. Einstein's Theory of Relativity provides the solution to the problem of whirling through history, at least for a passive form of time travel. A spaceship flying at almost the speed of light will be transported into the future according to Einstein's rules. The spaceship does not need a fictitious time tunnel, such as that seen in films, but exploits the effect of time dilation. As a result, time on board passes more slowly than on Earth. The crew therefore ages less quickly. On Earth, history continues as usual. Natural disasters leave behind destroyed landscapes. Economic crises destroy jobs and prosperity. To ignite wars. People live and die. Generations come and go. The spaceship does not take a short cut through space-time but waits until the required time has passed on Earth. At high speed or in the presence of strong gravitational fields, a waiting time, just as before the train transfer is sufficient. A few minutes. Several decades will have passed on Earth, during which time the spaceship crew would have aged only a little.

But what happens when the expedition arrives at a distant star and the crew suddenly decides to stop the undertaking and return home to their loved ones? In the meantime, years or decades will have passed on Earth and the relatives may not even recognise one another. Quite apart from the few shared memories and moments they can share after their return. Can the spaceship crew ever return to 2018? Can it reverse the effect of time dilation? Is the time dimension a "one-way" street or is it possible to turn the tables and get a return ticket? Can time travel be reversed?

The general theory of relativity is the starting point for deriving, proving or disproving assumptions about temporal phenomena. Even a hundred years after its publication, science has found no way of progressing beyond Einstein's theory of relativity in this respect. Einstein had discovered fairly early on that the theory of relativity would allow time travel. At least theoretically. But what is theoretically possible in practice is usually only a matter of time. However, since time travel could lead to irreparable causality violations, Einstein suspected that there was an unknown mechanism that would in principle prevent time travel into the past. Basically you travel into the past by exceeding the speed of light. An accelerated movement slows the time to the point where the speed of light is reached. Although this is actually impossible for particles having mass, as more energy is needed for each additional acceleration, if something reaches the speed of light its time stands still. If the object even succeeds in exceeding the speed of light, it will move back in time from then on. Particles moving at supraluminal speed, on the other hand, require an infinite amount of energy to be slowed down to the speed of light and fly into the past faster and faster as the energy decreases. However, the theory of relativity basically prohibits exceeding the speed of light, but of course it allows the existence of particles that are constantly moving faster than light. Such particles are called tachyons.

Tachyons always fly at supraluminal speed, moving backwards through time. Tachyons need energy to slow down to the speed of light, but they can never reach the same level as ordinary particles that have mass. Conversely, the less energy they have, the faster tachyons move The tachyon has quite strange, distorted features compared to the matter we know. However, we should not draw hasty conclusions, especially when it comes to questions of energy conservation or a possible Perpetuum Mobile that could arise from the use of tachyons. Tachyons cannot interact with ordinary matter. They elude the access of matter, moving at infraluminal speed. In addition, these supraluminal particles have an imaginary rest mass, so to speak, a negative mass (if you squared the imaginary mass). Therefore, the conclusion that an energyless tachyon moves infinitely fast and therefore has infinitely kinetic energy (kinetic energy) would be relatively meaningless. Moreover, our understanding of logic in the world of tachyons is an old chestnut. Since these particles move backwards in time, causality depends on the reference system. No one can therefore objectively judge which event caused which event. Tachyons are therefore to a certain extent the physical embodiment of the chicken and egg problem.

The theory of tachyons arose for a similar reason as the theory of antimatter, which later proved to be correct. For the equations of the theory of relativity show that exceeding the speed of light would not be possible or would be associated with an infinite energy requirement. The theory of relativity, however, allows for the existence of particles that always move at supraluminal speed. These particles arise as a mathematical solution of formalism. theory of everything also contain such supraluminal particles in some models. If they exist, it would mean that information could be sent into the past in any case. This statement provides an important key to the possibility of time travel back in time. Since we have never met time travellers from the future, we can make one or two hypotheses from the empirical viewpoint. First, our "time" is actual reality in this sense. There is no future that has already happened. Since we represent the "top of Time peak", we cannot have met any visitors from a future that does not exist yet. We cannot know the lottery numbers before the draw. Second, if past, present, and future exist in parallel with each other and the year 2500 has been the same as the year 2000, there is no way of travelling back in time, or no one has travelled back that far (or they have not made themselves known). Another way to explain the absence of time travellers could be that time travel is only possible through a kind of "train station". A machine that "welcomes" time travellers. Accordingly, time travel from the future into our present would only be possible once we had an appropriate infrastructure. This would only be the case if we had discovered and developed the fundamentals of this time machine.

To date, we have not discovered a fundamental principle of Nature that fundamentally prohibits time travel into the past. The second law of thermodynamics states that the disorder in a closed system will increase in time. If you drop a plate to the ground it will shatter into small pieces. Even if you wait a long time, the plate will never be put together again. So far no contradiction has ever been observed in this law, and indeed such a case would be quite momentous because then excavations and fossils could have arisen by accident and lose their historical significance. According to the law of thermodynamics, processes in Nature are completely reversible and hence also follow a backward-directed course of time. Not quite so certain is the situation with the weak and strong force, two basic forces of nature, since here the coincidence appears to determine the decay of atoms, so this cannot possibly be reversed. In the chapter on the glimpse into the third millennium we will apply the second law of thermodynamics to the question of whether the existence of Earth and life could have happened by chance.

We can already look into the past today. All we have to do is look at the starry sky in the evening. The light from some stars has been travelling hundreds or thousands of years to reach the Earth. It is possible that some of the stars we observe in the sky no longer exist. If our sun exploded in a supernova, we would not register this cosmic super-GAU on Earth until about eight minutes later. The light from each star gives us information about the universe of the past. With the best telescopes we can venture an ever deeper view into the universe. However, we can never say with certainty whether the galaxies and star system that have been discovered far away still exist. If the suns in the centre of the Milky Way crashed into a black hole in 2018, we would not be able to observe this cosmic annihilation about 27,000 years hence. That is how long the light would takes to reach Earth.

Incidentally, the speed of light is not exceeded if a spaceship A and a spaceship B move in opposite directions at 99 percent of the speed of light. In the theory of relativity, velocities cannot simply be added in the classical sense. Even at low speeds, it is actually erroneous simply to add the speeds together. Two cars travelling in opposite directions at 100 kilometres per hour do not move at exactly 200 kilometres per hour (the deviation is minimal at such low speeds). To determine the effective speed, a relativistic formula must be used at high speeds. Accordingly, two particles that move in opposite directions at almost the speed of light are still flying at infraluminal speed.

On the basis of the general theory of relativity, we know of certain possibilities of projecting us humans into the future, at least theoretically. But we do not know a viable way to undo this process, to move people into the past, even if the way into the past is not fundamentally blocked by the natural laws known today. However, the past may be final and Nature forbids travelling back in time to prevent anomalies and paradoxes. That would also mean that time is a one-way street with no possibility of turning back. Maybe we do not a sufficient understanding of the nature of time. This view might also be shared by Kurt Gödel, who has developed a slightly different theory on time travel. To the great regret of Albert Einstein.

2.2.9 Gödel's formula: time at the end of the universe

Kurt Gödel was an influential and prominent contemporary of Einstein. He loved mathematics and was interested in physics, especially the theory of relativity. He turned mathematics on its head when, in 1931, he formulated a sentence that proved its own unprovability. He loathed the public and was a quiet and withdrawn thinker who always presented himself correctly in a suit with neatly combed hair. On the contrary Einstein, who did not wear socks ("they only make holes anyway"), loved a confrontation with the public and looked more like a wild owl rather than a superstar of science.

Einstein and Gödel, two big friends, united by their common love of mathematics and physics. Two characters who could not be more different but who found common ground in the world of formulae and numbers. Hardly any other scientist of this time could hold a candle to either gentleman. Einstein also said: "I'm just going to the office for the privilege of being able to go home with Kurt Gödel." The formulae of general relativity were discussed for several hours during the walk home.

When Gödel was naturalised in the US, his friend Einstein accompanied him to the courtroom. When the judge read that Gödel, who had fled Austria at the beginning of the Second World War, was now in a free country where no dictatorship could ever break out, Gödel intervened energetically and denied it. His razor-sharp mind had noticed a logical contradiction when reading the constitution, which made the overthrow of democracy possible. Einstein, struck by the surprise element of Gödel's offensive, tugged at his sleeve and persuaded him not to continue the discourse he had initiated. The judge just shrugged, allowing for the fact that scientists were like that, and in the end they were all to become good Americans. But not only did Gödel make an impression as an extravagant figure, he shunned the public and married a nightclub dancer. Above all he excelled as a scientist. More than was to Einstein's liking. Gödel, for example, came up with the idea of drawing a conclusion from the formulae of General Relativity that Einstein did not want to support. To make matters worse, this momentous formula was to also be his seventieth birthday present. In short, Gödel's formula proves that time travel is possible and. above all, shows how to travel to any time. Einstein, who could not actually acquire a taste for anything that extended beyond our everyday imagination, apart from the theory of relativity, must have almost fell off his chair when Gödel presented him with his latest insights.

Until that point, pretty much all physicists considered the universe a sphere, which was obvious since stars, planets and finally the Earth are at least spherical. Gödel rejected this model and instead tried to calculate general relativity using cylindrical coordinates. He assumed that the universe revolves around an imaginary axis and is therefore constantly in motion. The galaxies and all matter are carried along by this movement. This creates the "Lense Thirring Effect", a twist of space-time. In the theory of relativity, space-time is known to be a four-dimensional structure. Each point is assigned four world lines or four coordinates. These world lines are entrained by rotation in Gödel's universe and are so curved that they eventually converge. The world lines are now self-contained. The surprising quintessence: If a spaceship flies are enough in the same direction, it does not reach the end of the universe but arrives at any point in time in the past. If you follow a world line curve long enough, you move backwards through history. Thus, the Middle Ages, the fall of Rome, the Stone Age or the Kennedy assassination are accessible. The catch: A world line curve does not become self-contained until after about 100 billion light years. A spaceship travelling at the speed of light would have to travel eight times longer than our universe has actually existed, a requirement that is difficult to meet.

Einstein showed himself to be reasonable and noted that the problem of time travel had already concerned him when developing the General Theory of Relativity. He suspected that there was a natural mechanism that would, in principle, prevent people from travelling back in time. Gödel, for his part, said that time travel is impracticable because of the astronomical lengths of a closed world-line. However, he forgot the Einstein-Rosen bridge, which connects two distant space-time areas and thus allows a kind of short-cut. Space-time is a structure of three-dimensional space and time. Thus a movement through space means a movement through time. An Einstein-Rosen bridge could be a short-cut through space-time, allowing much faster progression on the world line. Under certain circumstances, this would make it possible to arrive at the end of the universe or a closed world line within a very short time. But the Einstein-Rosen bridge is not the only way to significantly shorten cosmic distances. There are numerous ideas and suggestions for taking advantage of space-time curvatures, gravitational waves and the properties of general relativity in spaceships to enable interstellar travel.

All these bold proposals have one common problem: they cannot be realized in the foreseeable future and are therefore hypothetical, at least for the time being. We cannot build a spaceship that avails itself of these phenomena to travel through time or push into distant galaxies. Not because it would be impossible in principle, but rather because we lack the sources of energy and technology. To build an Einstein-Rosen bridge we need strange or exotic matter. In order to generate sufficiently strong gravitational waves or even to noticeably bend space we need an extremely large source of energy. One of the greatest challenges to the technical exploitation of the general relativity theory is to find a completely new type of energy. An energy form beyond fuels. An energy form that might be found in the field of quantum physics or nuclear physics. For example, by a reaction between various natural elements that releases antimatter. But maybe the path will lead us in a completely different direction which has been closed to us so far. A direction that revolutionises our energy potential, as it did over 75 years ago with the discovery of nuclear fission. Basically, time travel and flights to distant galaxies are probably not possible. Finally, time travel does not raise more logical problems than, for example, the widely accepted Big Bang theory. They are only impossible based on the technology of our time.

2.2.10 The black hole

The quantum and relativity theories have contributed significantly to our understanding of Nature over the last hundred years. They form the foundation and basis of modern physics. The quantum theories explain how the smallest particles move and behave. The theory of relativity explains gravity and thus essentially macroscopic systems. But does this actually do justice to the science of theories? You might well think so.

For under the cover of darkness a puzzling structure then emerged. A structure in which all previously known laws of Nature seem to strike. A structure that no human has ever seen. A structure that was already speculated by the British explorer John Michell and the French astronomer Pierre Simon Laplace in the 18th century. They wondered if there is a dark star somewhere in the universe. A star whose escape velocity is higher than the speed of light. A star whose surface no ship can ever leave, even when flying at the speed of light. The dark star was forgotten. Until Albert Einstein published the general theory of relativity and the German physicist Karl Schwarzschild succeeded only a year later in calculating the size and behaviour of a dark star. The dark stars or black holes, in popular parlance. They extend over huge areas in space. They reach for all matter, every planet, every star, no matter how big, and make them disappear into the void. Their gravity is so high that not even light can escape from the clutches of the black hole. There is no escape from the tentacles of this extreme cosmic phenomenon. Never has a human being seen a black hole. Its existence, however, results from a special case of the general theory of relativity. A theory that today has been confirmed experimentally and has remained largely undisputed. Model calculations in 1939 also revealed that the collapse of big stars requires a black hole. The clues that point to these cosmic destroyers are numerous. Astronomical calculations and observations confirm the suspicion that they must exist. And not just somewhere at the end of the universe, but distributed throughout the space-time. There is probably also a super-massive black hole in the centre of the Milky Way (with up to 3.5 million solar masses!) which greedily devours all matter that comes close to it. Although we can look deeper and deeper into space with modern telescopes, no one will ever see a black hole. It escapes our gaze at all times and in all places because its enormous attraction cannot even escape light from a certain distance. This critical boundary is called the "event horizon" or the "Schwarzschild radius". No light, no electromagnetic wave, no information that ever exceeds the event horizon, can leave it again. That's why we cannot see black holes directly. The enigmatic core of this cosmic entity, or the question of what a black hole actually is, remains veiled in absolute darkness to any outside observer, which is the reason why black holes can be detected with telescopes. When a black veil suddenly opens in the midst of a brightly shining sea of stars, we know that a black hole is eating up matter here. The space-time curvature at the event horizon is already so massive that it cannot even escape the light. This makes it impossible in principle to look behind the veil of a black hole without crossing and without returning. All matter that approaches a black hole sooner or later disappears into the cosmic vortex of uncertainty. Nobody knows what happens to matter there.

According to current knowledge, black holes have their origin in the death of very massive stars. After a few million or billion years, most stars have used up their burning stock. The celestial body begins to die. Nuclear fusion stops. What remains inside is a massive core of heavy elements. Now, no further nuclear fusion is possible. The radiation pressure, which has hitherto counteracted the space-time curvature (gravity) emanating from the very large mass of the star, disappears without substitution. The star collapses under its own gravity into an extremely compact celestial body with a highly increased density. If the star is lighter than about 1.44 solar masses it collapses into a white dwarf, a comparatively small star wreck. It does not develop into a black hole because quantum mechanical phenomena can build up sufficient counterpressure to gravity. If the mass of the dying star is between 1.44 and around 3 solar masses, the quantum mechanical pressure is too small to sufficiently counteract gravity. In this case, protons and electrons are compressed in such a manner that they lose their typical identity and become neutrons. This particle sump creates an extremely dense neutron star with a diameter of only a few kilometres. A neutron star is an extreme cosmic phenomenon which is only one step ahead of the formation of a black hole. A neutron star has a temperature of up to 100 billion Kelvins inside. The magnetic field and density also reach unimaginably high values. If the dying star has a mass of over 3 solar masses at the time of extinction, none of the other three basic forces of nature is strong enough to contain the gravitational effect. The star collapses. It collapses into a hypothetical quark star or a black hole - the most extreme phenomenon that has ever been observed in our universe (though only indirectly, as black holes are known to be invisible).

The resulting black hole bends space-time to such an extent that whole planets, stars, and galaxies are torn apart and face cosmic extinction. But what happens behind the event horizon? What happens to all matter? Where does it go? What happens at the core? What secrets are hidden behind this strange cosmic phenomenon? Standard Latin, when used in connection with 21st-century physics, is sometimes rather clueless when it comes to explaining this "voracious" behaviour.

The only certainty is that a black hole causes a strong space-time curvature and thereby attracts all matter (we remember: the range of gravity is - according to current knowledge - infinite.) This could be with the discovery of the "graviton", a hypothetical Transmission particle, possibly change).

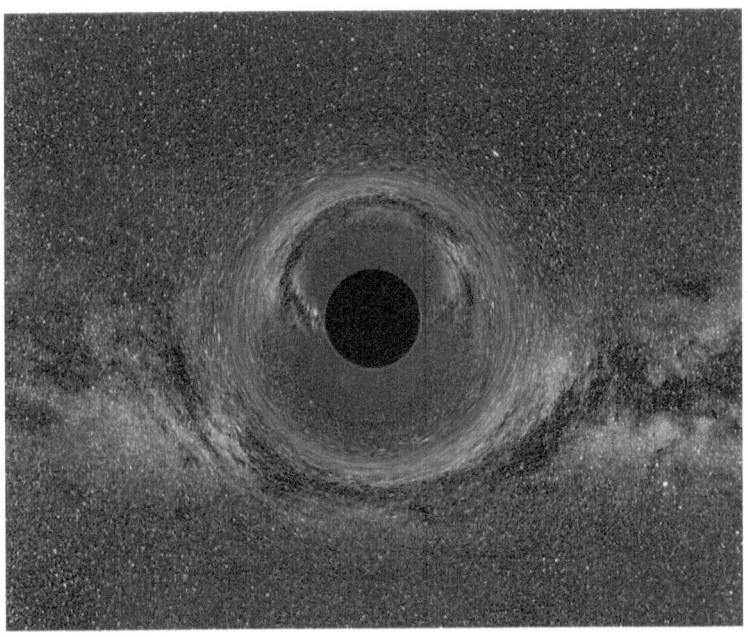

Figure 5   A black hole from a distance of 600 km

The Strange Universe: Einstein, Quantum Physics and the ToE

In order to escape the fictitious black hole in Figure 5, you would need a spaceship steering in the opposite direction with over 400 million times gravity acceleration. It would take 400 million of our planet to cause the same gravitational force as the black hole.

Some rather conservative scientists hold the view that the entire mass that a black hole absorbs during its lifetime is concentrated at a single infinitely small point, so-called singularity, which is supposed to arise from an almost infinite curvature of space-time. At the same time, the validity of the natural laws known to us is once more effectively suspended. Just as rational comprehensibility. The mathematical prediction for the processes that take place in the black hole is limited to the concentration of an arbitrarily large mass at an infinitely small point. Conditions that are assumed to have prevailed right at the beginning of the universe.

Usually, physicists regard infinities in formalism as a sign of inadequacy due to a lack of knowledge or a mistake in theory. This is also the case here. The black hole is an extreme phenomenon that pushes the General Theory of Relativity to the limits of its validity. For the processes assumed therein overlap macroscopic and microscopic effects. On the one hand, there is a huge space-time curvature that falls within the territory of general relativity. On the other hand, in the core, at supposed singularity, quantum-mechanical processes would have to take place. From this it is already possible to recognise the problem with which the scientific world has to contend: In order to explain phenomena in which extreme gravitational phenomena encounter microscopic, quantum mechanical phenomena, a superordinate theory, a combination of the General Theory of Relativity and quantum physics, is indispensable, for good or evil.

Nonetheless, it is possible to come up with theories about the inner workings of black holes without pulling the theory of everything out of a hat when we open the horizon and are ready to enter a new sphere that may revolutionise tomorrow's world view, as Einstein did in the case of today's world view. Let's take another look at the processes that lead to the formation of a black hole: A massive, burnt-out star dies. The inner jet pressure is drying up. Gravity overcomes all other basic forces and causes the star to collapse, to fall in on itself. The entire mass is concentrated in a very small space. The resulting density is enormous and continues to grow with increasing compression. This process leads to extreme states that we have never experienced in everyday life or experimentally. There are now three possibilities with which to describe the subsequent events. First of all, the third possibility is an approach based on string theory, a potential theory of everything currently under development. This approach will be explained separately later. The first option takes us back to the path the ambitious pilot took with his spaceship. He accelerated the spaceship ever faster and faster until he had crossed the Planck boundaries. The same principle could underlie the formation of a black hole. The enormous mass, which collapses inexorably, is increasingly compressed by the reinforced space-time curvature. The process builds up and eventually reaches a dangerous momentum as more and more mass is squeezed into an ever smaller space, giving rise an increasingly extreme curvature of space-time. At some point the space-time curvature reaches a critical level and finally strives towards infinity. This creates conditions such as those assumed to form singularity or generally to prevail at the core of a black hole. The mass of the entire star is now concentrated in the smallest space. Planck density, i.e. the maximum density that a body may assume without jeopardising the stability of space-time or calling into question the validity of the laws of nature, is exceeded. This creates a crack[31] in space-time. structure. The space-time curvature reaches a degree that the event horizon may assume for a few kilometres in diameter. As a result of this interpretive variant, the tangled matter does not concentrate at a mysterious point that cannot be conclusively explained by modern physics, but is torn in one pull from space-time. The cosmic hole is preserved as long as there is an event horizon. Gravity is therefore so strong that at a certain distance even light cannot escape. Nobody knows the exact lifetime of a black hole. Stephen Hawking postulated in 1974 that black holes give off so-called "Hawking radiation". In some cases, the emitted radiation will exceed the attracted mass, causing the black hole to "evaporate" sooner or later.

This theory could contribute significantly to the fate and future of humanity in just a few years. It is not that the black hole in the Milky Way is developing into imminent danger. But scientists are currently conducting controversial experiments at the "LHC" particle accelerator in Geneva, where heavy lead atoms are bombarded with very fast protons. Physics hopes to deduce from this the conditions that prevailed at the time of the Big Bang and to discover new particles that have already been postulated in any case. These experiments are very energy intensive and could lead to the formation of black mini-holes or strange matter on Earth, for example when string theory, a contender for the theory of everything, is correct to some extent and wound-up dimensions

---

[31] The extent of this crack would probably only be small, but quantum mechanics cannot adequately describe such events. An overarching theory, which is possibly just being discovered in the form of string theory, is probably required here.

actually exist. In the opinion of leading experts, these mini-holes should decay shortly after their formation due to their small size. At least one is an assertion because no-one knows this exactly. Experiments such as those at the LHC have never been done before and the theoretical foundations are too inaccurate to predict the behaviour of a black mini-hole. It would only be really dangerous if it turned out that there is no Hawking radiation in Nature. In that case, at least in the long term, there would be a serious threat. The black mini-holes would settle in the earth's mass and slowly devour the entire planet within a period estimated at between 50 months and 50 billion years (!)[32].

If string theory is correct, black holes would have to emerge at least once a year in the Earth's atmosphere through the impact of cosmic rays. However, this cosmic radiation travels at such a high speed that the initially extremely small black mini-holes roar through the earth into the vastness of the universe and thus pose no threat to our planet. Science postulates that in the "LHC" accelerator one black mini-hole per second could arise. Two American citizens found the matter altogether too much. They filed a lawsuit against the commissioning of the "LHC" in a US court. This incident documents how controversial and feared these particle experiments are to some extent. Although some universities are individually firing live ammunition at the experiments being conducted at the "LHC", the effects of the experiment are only half as dramatic as assumed. A cosmic catastrophe cannot be ruled out with absolute certainty. The mood of some scientists, however, serves more to heighten the drama and their own need to be in the headlines than to promote objective

---

[32] These estimates show how little science knows about black up till now. The predictions for the "worst case scenario" are just as imprecise and manifold.

debate. The tabloid press and hunting magazines rarely back away from doomsday scenarios, particularly when the headline feed comes from more or less "reliable" sources. With all due caution one should not forget that the emergence of an imminently dangerous black hole represents an extreme cosmic phenomenon, which alone would be impossible to reproduce in terms of energy. Small mini-holes may take several million years to grow to a threatening size. An unpredictable situation could at best arise if a process produces strange matter that is normally inseparably bound up with ordinary matter, as we shall see shortly.

What happens to the matter that crashes into a black hole? The crack in space-time has a similar effect, comparable to a filled bathtub where you pull the plug. The water in the tub flows through the drain, developing a noticeable suction. Similarly one can imagine the processes that take place inside a black hole. Matter is sucked through the crack and transported through this crack from space-time. What is interesting is the question of where this suction leads. The water in the bathtub flows through the sewage pipes of the building and into the sewage system. But where does the matter in the black hole go? Perhaps this cosmic one-way street leads to the postulated hyperspace, the five-dimensional universe that surrounds our space-time. Unlike a single object (for example a stone or rock), which overloads space-time once and eventually ruptures it, the critical limit in a black hole is constantly or even permanently exceeded. At least until it becomes contaminated or otherwise dissolves. The unimaginably strong gravity continuously attracts new mass, which continuously overloads or stresses the space-time structure even more. This opens up a permanent tunnel of unknown size, which is fed and maintained with the attracted mass. The suction in the bathtub exists until the plug is re-inserted

(= the space-time gap closes) or the bathtub (= space-time) is empty. The mass, which in the black hole disappears into thin air, enters another universe or a remote area of space-time. Contrary to the theory of singularity, which provides for a concentration of unimaginable masses at a single, one-dimensional point of the singularity. It would even be conceivable that the creation of the black hole creates and expands a new space-time, a parallel universe. In this context, the black hole in the new universe appears as a white hole, which ejects and repels matter and thus develops an anti-gravitational effect.

Depending on the interpretation, the black hole can be understood as a hole in the sphere of space or as a tunnel and escape route into another, strange, unknown world. As a stepping stone into hyperspace or to see what is really happening, but a black hole is not recommended. Gravity is so strong that each person would simply squish and at the most would land as a tiny particle in hyperspace.

One interesting question which comes up again and again relates to the event horizon, the "point of no return", the point at which it is no longer possible to run from the gravity of the black hole. In this area the gravity is so strong that time practically stands still according to the gravitational time dilation of the General Theory of Relativity (purely theoretically the gravitational field of the event horizon could be used as a time machine. But what happens beyond this horizon? Does gravity exceed even this maxim? Does crossing the event horizon even mean a negative acceleration in time, a negative passage of time, which could be equated with a movement into the past? Or is the event horizon limited to sketching the rift of the space-time. structure? And what happens to the mass that flows into hyperspace?

Questions about questions whose answer lies in the stars. We recall that phenomena like the black hole exceed the Planck constants and therefore cannot be grasped in principle on the basis of our laws of nature. Perhaps the black hole in hyperspace appears as a white hole from which matter arises. If our universe is a multi-dimensional space-geometric construct that somehow merges with another one (similar to the antagonism of the ball), it would even be conceivable that a black hole represents a link to another point in the universe. This would also solve a problem that divides the world of experts until today. A basic principle of physics states that neither energy nor mass nor information can be lost easily. How can this principle of energy conservation be reconciled with the whimsical nature of the black hole?

As already hinted, Hawking radiation provides a solution to this. Accordingly, black holes emit information and energy in the very same radiation and thus present a possibility that matter may again escape from the loops of the black hole. On the other hand, the linking of a black hole to our or any other universe is an explanatory approach. If the black hole absorbs the information and releases it elsewhere or in another universe, it would only be transferred from one space coordinate to another, possibly a very distant space coordinate. The explanation depends, in turn, on the objective structure of our universe or the question of how our universe looks from the "outside." Is there only our universe or are there numerous parallel universes in which our universe is perhaps embedded as one of many? In this case, it would also be conceivable that the conservation of energy is universal, that is, considered across all universes. Information, matter and energy would not be destroyed in the black hole, but merely spatially shifted. In roughly the same way as the drain makes the water of the bathtub disappear, but does not destroy it. It just flows into the sewers. The conservation of mass in your bathroom may be negative, but throughout the Earth, of course, no mass is lost. The same could apply to the black hole and our universe if we assume that the matter sucked in does not just disappear but reappears somewhere. Even though this somewhere may not be accessible to us.

A second, more speculative theory does not necessarily assume a rupture of space-time. in the core of the black hole. Rather they are decomposed by the strong gravity, the elementary particles in their nearest constituents. The elementary particles are composed of so-called quarks, one of which deserves special attention: the Strange Quark. It is called strange because it is not created by the same force by which it decays. A neutron contains two "down" quark and one "up" quark. Some physicists now assume that there are conditions in a neutron star that break down neutrons into their constituents, turning a "down" quark into a "strange" quark. What sounds cryptic and complicated is extremely significant. The "strange" quark differs significantly from the other quarks in terms of another property: If it were to succeed in producing stable, strange matter, it could produce an attractive effect similar to that of a black hole.

Strange matter is supposed to attract and absorb normal matter[33].

Already from a mass equivalent to a thousand protons, weird

---

[33] This force of attraction emanating from strange matter cannot be explained by the four known fundamental forces of physics. Is the key to a fifth fundamental force, for which some physicists have hitherto searched to no avail because of the incompatibility of the theory of relativity and quantum physics, to be found in strange matter?

matter might be completely stable. Calculations lead to strange matter hardly being able to be generated in the particle accelerator. However, stable strange matter might exist in a neutron star or in an even more extreme black hole. For there exist extreme states, the explanation for which quantum physics and theory of relativity would have to be integrated in a more comprehensive theory. It would even be conceivable that in the nucleus of a black hole such stable strange matter led to the absorption of the attracted matter. Or would even be responsible for part of the attraction. When looking at black holes, the problem is that the space-time curvature already develops an acceleration (gravity) at the event horizon that is stronger and, in principle, faster than light (otherwise the light could escape). In principle, however, such a gravitational effect is a space-time curvature that is incompatible with the four-dimensional space times. This is a space-time curvature that causes a sink with an almost infinite slope and is accompanied by a spatial "destruction" of the space-time structure. It might be conceivable that gravity is limited to a tolerable (but nevertheless extreme) measure, at least outside the centre. The phenomenon that even light cannot escape from the horizon of events would then no longer be due solely to the gravitational effect but to a significant hitherto fifth (hypothetical) fundamental force of physics. A fundamental force that has remained undetected so far but which has at least not been excluded by some scientists. A basic force that can be found in stable, strange matter.

2.2.11 Antigravity and dark energy

In physics, as we know, there are many gaps left to fill. So one assumes that for every physical existence a counterpart exists. And for every plus there is a minus, for each force a counterforce, for

The Strange Universe: Einstein, Quantum Physics and the ToE

each particle an antiparticle. In macroscopic spheres, in the dimensions of the solar system and the universe, this symmetry principle is not far wrong. If there is a counterpart to any physical existence, then there must be a counterpart to gravity, energy, or the black hole, at least when nature is consistent and exhibits the same behaviour as the band, no matter whether we are in the world of the small (quantum physics) or in the world of the big (relativity).

But where are the counterparts in the world of the great? Where do we find the counterparts to gravity, energy or the black hole? And how fundamental is the symmetry principle really? Is there symmetry with everything in our universe or even with the universe itself? A kind of mirror universe? And what about dimensions, space or time?

If we take a look at the General Theory of Relativity we can assume that there is a counterpart to gravity, antigravity. To illustrate this, gravity is a sink in space-time, a kind of "crater" into which objects slip, creating what we call gravity. Antigravity can be imagined as a hill in space-time. All the objects that are on this hill slide down. The stronger the antigravity, the higher and steeper the hill. Antigravity always acts as a repulsive force.

The million Euro question[34] now to be answered is whether such antigravity exists in Nature at all - after all, it has never been observed - and how such antigravity can be generated. The General Theory of Relativity is not very visionary on this fundamental

---

34   The "Göde" Foundation is awarding a prize of one million Euros for the first reproducible experiment to "overcome gravity". If you succeed in effectively causing a mass to float for one minute through the influence of gravity, you will be richer by a lot of money and a Nobel Prize.

question. According to Einstein, at least the shielding of gravity should not be possible. For this would violate the principle of equivalence whereby all energies and masses go through the same fall curve if the starting point and the speed coincide. A ball, a spring and a hammer fall in the vacuum exactly equal to the ground. The fall curve is independent of the nature of the matter. However, if it were possible to shield gravity, the fall curves could be manipulated and the equivalence principle violated. However, this principle has been clearly confirmed in all experiments conducted so far. If an experiment succeeds in refuting the principle of equivalence, this would be an indication of the incompleteness of the General Theory of Relativity and hence a strong argument for the existence of a higher-level theory that has been suspected by science for decades. In the light of experimental evidence, the shielding of gravity seems to be one of the routes into a core principle of relativity. But what about the existence of antigravity? The sword of Damocles for Einstein or just fantasy?

Antigravity could well be compatible with the theory of relativity. At least it would not violate the principle of equivalence in that at the same place and speed all energies and masses would continue to go through the same fall curve - even if the falling curve does not lead to the centre of the source of antigravity but repels the objects. However, as with gravity, the fall curve would be the same regardless of the nature of the objects. On the other hand, one could of course argue that antigravity might also serve to isolate or weaken , but this in turn need not be a contradiction because within the resulting total gravity all bodies again pass through the same fall curves.

From our experience, we only know the gravity that holds us on earth and significantly shapes the structure of the universe. In order for antigravity to arise, space-time does not curve, but rather "rises," a counterpart to mass or energy is necessary. Here we are not talking about antimatter, because antimatter is only the counterpart of matter in terms of particle charges. The protons, electrons and neutrons of antimatter just have the opposite charge to the protons, electrons and neutrons of matter. In addition, according to Einstein's famous formula "e = mc2", energy is equivalent to mass. When matter and antimatter meet, they radiate energy. Every energy in turn bends space-time. However, since we want to create a hill, not a sink, in space-time, we obviously need the counterpart of energy to create antigravity.

But what is the counterpart of energy?

It is matter with negative energy density, which is pithily referred to as exotic matter. Exotic because this matter cannot consist of protons, neutrons and electrons, but also because matter with negative energy density is beyond our understanding and has not yet been observed. The confused notion of antimatter in the minds of people who do not know exactly what antimatter is, is therefore more likely to apply to exotic matter.

With exotic mass it would be possible in principle to develop an anti-gravitational effect in which space is not curved but, as it were, inflated. So far exotic matter only exists hypothetically. No one knows if it even exists. Perhaps space-time curvatures are only possible in our universe, so there can only be one attractive force, gravity. However, it is conceivable that in surrounding hyperspace or a parallel universe the space-time curvatures act as space-time elevations and thereby develop a repulsive force, an antigravity. Just like a stone on a taut aluminium foil creates a sink when viewed from above but looks like a hill from below. Another indication of the existence of antigravity is the postulate of the White Hole, the opposite of the black hole. A white hole ejects matter and is thus to some extent the embodiment of anti-gravity par excellence.

One form of antigravity is the hypothetical dark energy that, according to conventional wisdom, occupies about two-thirds of the universe. Dark energy is used as an explanation for the experimentally determined accelerated expansion of the universe. Due to gravitational forces a deceleration of the expansion would have been suspected, but whether the cause of this accelerated expansion is ultimately dark energy - and what dark energy is supposed to be - is the subject of current research. On the one hand, it is believed that dark energy could be due to the vacuum energy. Because of the uncertainty principle of quantum mechanics we know that nothing should be at absolute rest. Even in a vacuum there would therefore have to be a so-called quantum fluctuation, a certain quantum mechanical zero-point or residual energy. This could be interpreted as dark energy. However, there is no quantum theory derivation that would convincingly explain this relationship. Perhaps dark energy cannot be conclusively explained in the context of current physics. It might even be conceivable that the actual cause is due to a parallel universe. Let us recall the hypothesis that space-time curvatures in our universe could develop antigravity in a parallel or surrounding universe. In this case, the accelerated expansion of our universe could be due to anti-gravitational forces from another universe. Although it may sound like illusions and far-fetched explanations, it is still a serious alternative worth considering. Especially when we consider that gravity, for reasons as yet unexplained, is by far the weakest of the four known fundamental forces of Nature. In an attempt to explain this, American physics professor Lisa Randall proposed a five-dimensional space-time. Our universe therefore consists of two boundary worlds separated by an empty space. In the one boundary world the three basic forces exist: electromagnetic, weak and strong interaction. Light and all elementary particles can only be in this boundary world in which we also exist, and cannot interact with the second boundary world. Gravity, a hypothetical particle that transmits the force of attraction can jump the gap and link the two boundary worlds. Gravity is therefore the only fundamental force that works through all dimensions and worlds, and because of this our four-dimensional space-time (one of the boundary worlds) is much weaker than the other three basic forces. Of course, this model is quite speculative, but compared to world theories it is modest in terms of dimensions and complexity.

The accelerated expansion of our universe, dark energy, could therefore be due to a gravitational effect from another world, which in our space-time acts as a repulsive force, causing accelerated expansion. Of course, this hypothesis is only one of many that are possible. The thrust of current research, however, is heading in the direction of models that at least consider a world with numerous hidden dimensions and universes to be possible.

An alternative explanation for dark energy could also be found in an inadequacy of the General Theory of Relativity. If this theory is only one limiting case of a superordinate theory - and thus must be assumed - the necessary adjustments could at most lead to dark energy disappearing in the course of an adapted formalism. After all, dark energy only arose to explain the differences between theoretical expectation and the actual observation of the expansion.

In addition to dark energy, there is also dark matter. According to the laws of physics, the rotational speed in the outer regions of galaxies would have to decrease as in the case of Pluto revolving around the sun much more slowly than Mercury, the planet nearest the Sun. In fact, observations show that the rotational speed remains constant or is even faster. A higher rotational speed allows the assumption of previously as yet undiscovered masses, resulting in a higher force of attraction and hence a faster rotation. These masses do not exist in the form of known matter such as stars, planets, gases or cosmic debris, which is why they are called dark matter. In addition to the existence of hypothetical particles, a modification of the General Theory of Relativity is considered is believed to be one explanation. Accordingly, it is postulated that the equivalence between inert and heavy mass at very small accelerations is no longer valid in any case. Perhaps dark matter and dark energy are due only to an inadequacy in our theories and not to hitherto undiscovered, puzzling phenomena. In any case the fact that 95 per cent of the entire universe must consist of hitherto undiscovered, unknown dark matter or dark energy does not exactly point to the completeness of a theory to explain the cosmos[35].

Once again, we have to admit soberly that modern physics can only explain a tiny piece of the universe. To really understand our Nature, we have a long and adventurous journey ahead of us. But first let us first venture into the world of the small, into the world of quantum physics. Here we expect far more bizarre, stranger and more explanatory phenomena than in the theory of relativity because quantum mechanics contradicts our everyday experience in

---

[35] Modern cosmology assumes that the universe consists only 4.6% of atoms, 72% of dark energy and 23% of dark matter. The remaining 0.4% consists of neutrinos, cosmic radiation, etc.

The Strange Universe: Einstein, Quantum Physics and the ToE

many ways and fundamentally changes our view of the world.

But sit back, treat yourself to a cup of coffee or a good cup of tea and enjoy the tranquillity that still reigns between you, this book and the couch on which you are hoping to recover. Because once you let quantum physics into the house, pure chaos returns and your couch will never be as innocuous as it is (hopefully) at this moment.

3. The quantum nightmare

Quantum physics describes phenomena and processes in the microscopic world, the world of the small. In this atomic and subatomic realm very different laws of nature prevail to those we know from everyday life. Laws that run counter to the principles of classical physics that are considered to be self-evident and fundamental. The particles do what they like and behave like a stubborn child. The rules coincide, at least - and this is one of the greatest mysteries of Nature - until Man intervenes and tries to measure the behaviour of the particles in an experiment. Then the quanta will bend the rules ascribed to the microscopic world and behave as we would expect from our experience.

The essence of quantum physics is very strange and fundamentally different from the principles of the theory of relativity or classical mechanics intuitively familiar to us. Until the discovery of quantum physics all processes in Nature were considered to be predictable, at least theoretically. A very powerful computer capable of accessing all the relevant information in the universe would therefore have been able to accurately predict all future physical events. With quantum physics, however, randomness moves into the house of science. The smallest particles behave mysteriously and unmanageably. They are everywhere, they move along several different paths at the same time and jump spontaneously into forbidden zones in which they cannot actually sustain themselves. The movement of the particles can only be estimated with probability calculations. It's like a shell game: as long as no one is looking, the ball can be everywhere. The elementary particles lose their distinct character and, thus physics cannot claim to be able to predict the course of events in principle. Probability, as a kind of internal protective mechanism of Nature, prevents events from being reliably predicted, independently of the experimental information available through a system. Determining the exact location or movement of a particle is therefore impossible in quantum physics. Not because our equipment or technical skills are inadequate but because it is a fundamental principle of Nature, which is puzzlingly true only in the world of small particles. In the everyday world known to us, everything continues as usual. Computable and consistent processes dispel the mystery of quantum mechanics. The reason for this fundamental difference between the microscopic and the macroscopic world has still not been unravelled and is the subject of the most recent research.

Quantum physics is the key to the most fundamental questions of the universe. But it also brings many new questions into focus. For example, questions about compatibility with the theory of relativity and the theory of the macrocosm, which seems, however, to depend on entirely different principles. How can it be that in the microcosmos and macrocosm such different laws of Nature prevail? Can these laws of Nature be combined under one common denominator? How can the mysterious phenomena of quantum mechanics be explained? What is behind the coincidence?

We will find answers to these and many more exciting questions and amazing phenomena from the world of the quanta in the following chapters.

The quantising of light, space and time

At the beginning of the 20th century quantum physics experienced its great pioneering age. Physicists such as Planck, Heisenberg,

The Strange Universe: Einstein, Quantum Physics and the ToE

Schrödinger, Dirac or Born developed the theories of quantum mechanics and therefore constructed, alongside the theory of relativity, the second pillar of modern physics. Albert Einstein made an important contribution with his light quantum hypothesis, but he could not acquire a taste for the mysterious form of quantum physics at any time in his life. Einstein was a journeyman who appreciated the beauty of Nature. He disliked whatever questioned the elegance and stability of the universe including, in particular, the recently discovered quantum physics, which introduced randomness into the house of Mother Nature.

An unwanted guest, found Einstein. But he knew it was he who had shaken up the essence of light with his quantum light hypothesis. And so he turned out to be an escort to this unwanted guest.

But first things first. What is this quantum nightmare and why does it change our view of the world so drastically? And how in the world can rational people come up with the idea of devising such a theory that opposes everything we know about Nature?

Around the year 1900, science had to contend with some strange inconsistencies that had arisen between theory and experiment. One sticking point was the so-called photoelectric effect, and with it the question of the century, what light actually is. The English physicist Thomas Young had already given a conclusive answer to this question in 1802 with his famous double-slit experiment. He had shown at that time that light overlaps like waves. Consequently, light had to be a wave since only waves can form such superpositions (interferences). In 1839, Alexandre Edmond Becquerel discovered the photoelectric effect: when irradiating a metal surface with light, electrons are struck from the metal (see

Figure 6). The problem here is that in theory completely different results were expected from those that could be reproduced in the experiment. No Hawaiian Pizza came out of the oven of physics, even though the cook had thrown in a heap of pineapple with ham and cheese. The dilemma of the photoelectric effect can be illustrated by means of a microwave. It was assumed at that time that light is a wave. Waves transmit their energy continuously, i.e. over a "longer" period of time. A prime example is your microwave oven. If you place your pasta dish in the appliance and then turn it on, the dish is not warmed immediately. It takes a while, usually two to three minutes. During this time, the microwave transfers energy to your food, which heats it up. However, the energy is not transmitted immediately, but constantly - that is, continuously - to the food. Likewise, light would have to behave in the photoelectric effect under Young's assumption that light is a wave. So when a metal plate is irradiated with light it should recharge and have enough energy to release electrons from the metal after about one second.

So much for the theory. Unfortunately for some scientists, it should be shown in the experiment that the photoelectric effect occurs immediately when the apparatus is switched on. The electrons are struck directly out of the metal and not just after a few seconds, as would have been expected due to the wave nature of the light. In addition, it has sometimes been suggested that the intensity of light has an effect on the rate at which the released electrons would fly out of the metal.

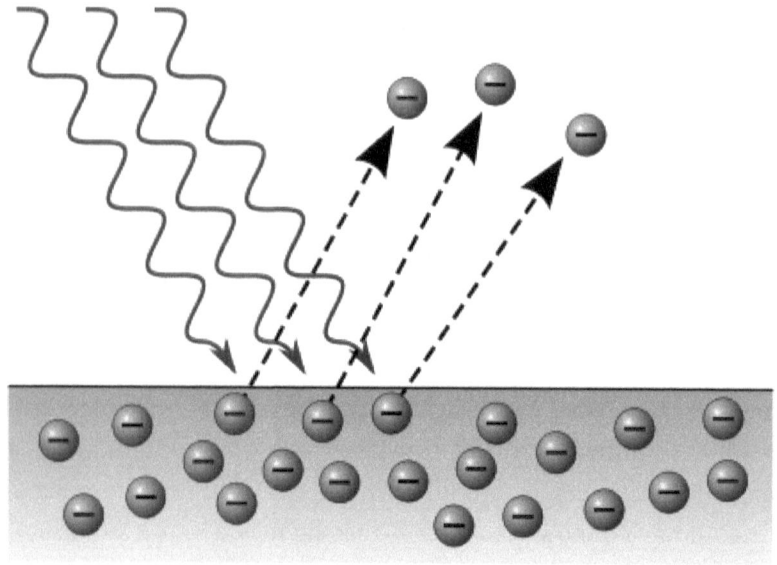

Figure 6   The photoelectric effect ("short-wave light")

The higher the intensity, the more energy is transferred to the metal and the more violent the electrons would have to be struck out of the metal. In the glistening midday sun at the Equator it finally "tans" much more intensively than in late winter evenings in front of the Alpine hut (except when the farmer is spraying liquid manure). But even on this point the experiment should teach the professional world something, no matter what the price, thus causing amazement and perplexity. No one had any idea of the far-reaching significance this experiment would one day attain. The photoelectric effect was one of the most remarkable contradictions of nineteenth-century theoretical physics. A dead end. A dilemma.

The Strange Universe: Einstein, Quantum Physics and the ToE

Then came Albert Einstein. He concocted a new recipe. He saw, came and conquered. Einstein explained the photoelectric effect in 1905 on the basis of a ground breaking idea of the German physicist Max Planck, an idea that was to change the world forever.

Max Planck had discovered in 1900, when researching the black body problem, that energy does not occur "all at one time" (continuously), but only in small packets. Each amount of energy represents a multiple of these basic packets. Science, and especially Planck himself, considered this idea to be a contrived simplification to ensure that his formulae made sense and coincided with the experimental findings. But Einstein recognised the meaning of this idea and consequently applied it to light. This resulted in his treatise on the photoelectric effect, for which he was to receive the Nobel Prize sixteen years later. Quantum mechanics was developed from the Planck's theory by Albert Einstein. In quantum physics it is simpler to quantise physical processes, that is to say, sort them into packets. This means that certain properties in Nature only exist as multiples of a very small basic quantity or basic property. Here it is said that these properties are "quantised". A prime example of this is the electrical charge. Contrary to what is assumed in everyday life, the electrical charge cannot assume any value but must always be a multiple of the charge of each individual electron. The electrical charge is therefore said to be quantised. This smallest electrical charge[36] may be imagined as a postal packet sent to a godchild for Christmas. This packet cannot be cut, split or halved. The more gifts you want to send the more packets you need. Consequently, the entire charge can only ever assume whole (= discrete) values. For example, 5, 10 or 15 packets, but not 5 ½ or 8 ¾ packets. This means that every charge, no matter how it is

---

[36] The smallest electrical charge is the charge of an electron.

obtained, is always a multiple of the smallest charge. An electrical charge can be 100 or 1000 packets, but never 3 ½ or 10 ⅞ or similar. One packet represents the smallest possible unit and therefore can never be split. Just like in an airplane 80, 100 or 120 passengers can fly, but never a fraction, for example 80 ½.

The first important statement of quantum mechanics is therefore that certain physical things exist only in quantised form, that is always a multiple of a basic unit. For example, the smallest Euro currency unit is the 1 cent coin, which is not a fraction but from which any higher cash amount can be composed.

The discovery of quantisation should was to be highly significant. Before Einstein published his treatise on the photoelectric effect, light was a wave. The double split experiment of Young had proved the wave nature of light. But Einstein explained that light can behave like a particle. Really. In considering light as a stream of particles[37], the exact deviations observed in the photoelectric effect between the theory and the experiment result are produced. Light is therefore wave and particle at the same time. In Young's experiment light displayed the nature of a wave, but under the photoelectric effect it suddenly behaved like a particle. But how can something be particle and wave at the same time? And how do we know if light is now a wave or a particle?

Einstein explained the photoelectric effect by assuming that light is a particle. Accordingly, light consists of many small packets of energy, so-called photons. These photons are massless and always

---

[37] The "particles" are the photons, the individual energy packets of light and the electromagnetic radiations respectively. Photons are without mass and always move at the speed of light. Newton did not guess that such a flow of particles existed but rather a flow with particles that have mass.

move at the speed of light. Einstein said that an electron is struck from the metal as soon as it is hit by a photon ("light packet"). Therefore, the photoelectric effect occurs immediately and not after the expected second. To illustrate this fact, think about a pool table. If you strike the white ball with your cue to sink one of the coloured balls into one of the holes, the white ball moves in the exact direction of the stroke and transfers its motion to the ball you hit. The entire billiard game takes place happens immediately. You would also look pretty confused from the laundry if that was not the case (provided that no beer pack was involved). But if the white ball were a wave, it would "continuously" transmit its motion on collision with the other ball. Just as microwave cooking takes a few minutes to get warm. Thus, the white "wave" ball does not sink the red ball immediately into a hole after the collision, but some eerie moments pass until the red ball starts to move. It looks like the balls were pausing in the collision before responding to the impact. Since the exposure of metal leads without any time delay to the photoelectric effect, the incident light must have the character of a billiard ball. Light does not behave as a wave, as one falsely believed, but as a particle. As particles, as Einstein postulated.

Einstein was able to explain the inconsistencies between theory and experiment by postulating that light can behave like a particle. So light is both a particle and a wave. Light, and of course every other electromagnetic wave, can behave as a particle as well as a wave. In Young's double-slit experiment light behaves like a wave as well as a particle in the photoelectric effect. This very strange phenomenon is called wave-particle dualism. Why light can behave simultaneously[38] as a particle or wave, and which mechanism

---

[38] Actually light behaves like a particle or wave – but until measured both behaviours overlap, which is why light is simultaneously a particle and a wave.

communicates to light how it should behave, is still not understood. This wave-particle dualism is one of the greatest secrets of modern physics - and it leads even further, even deeper into the jungle of the quantum world. If the double-slit experiment is conducted with particles of matter, electrons for example, the result depends essentially on whether the electrons are observed or not. Depending on that, they appear as waves or as particles. You see, the world of quantum is eerie. And it's huge.

Einstein subjected the electromagnetic radiation and thus light to the verdict of quantisation. In doing so, he destroyed the notion of the consistent world we experience in everyday life. We consider a light beam from a commercial laser pointer as a continuous red line. Likewise, we have the feeling that the electrical current flows continuously from the socket. Anyone who has come into unpleasant contact with a cattle fence knows what I mean. Once again these are fallacies that seem to be correct in everyday life but in fact are only approximations of reality. Neither light nor electric current flow continuously, but are quantised. Light is divided into small energy packets, called photons. A beam of light is nothing but the stringing together of many such energy packets. The lamp in your room does not emit light constantly, but packet by packet, photon by photon. If you place an order with an Internet shop you will also receive the delivery "quantised". The postman hands it to you packet by packet. Each packet represents a quantum, the smallest unit that cannot be split, halved or otherwise divided. When you turn on the lamp, electrical power is drawn out of the socket packet by packet and light is radiated from the bulb packet by packet. Of course these packets are extremely small, which is why we think that the current is flowing continuously and the light bulb is emitting light constantly.

However, we could slow down the light drastically and watch it in slow motion[39] and realize that light is not a coherent ray but consists of almost innumerable small packets of energy (photons). This revolutionary insight cemented the foundation of quantum mechanics [40]. Numerous inconsistencies and problems could be explained by this completely new approach to physics. As with any new discovery, however, many new unknowns have emerged.

3.2 The Eerie World of Quantum

Quantum physics is an eerie peak castle with many trapdoors, locked rooms and ghostly skills. A stony path leads along the cliff up into the misty landscape of this building. No signpost discloses the branches. Nobody knows where the road leads. All the bold pioneers have lost their way in the thicket of this world. This insight is not reached until the one crosses the moat, enters the castle gate and casts a glance into the frighteningly fascinating rooms. The fireplace crackles in the big hall and a shadow is cast over the walls. A half-dead cat wraps itself around the chair leg. From the porcelain bowl on the heavy wooden table jumps an apple. Between two thin slits falls a light that no one can ever see because it disappears immediately when someone is looking. The chair marches into the fire. And suddenly there is a suit of armour in the hall. Empty, of course. A gloomy scenario from the world of

---

[39] Although in practice this would not work because of the constancy of the speed of light, it will serve to illustrate this.

[40] In the case of the photoelectric effect and Planck's discovery of quantisation in 1900, reference is also made to "classical quantum mechanics", since the knowledge had only been applied to physics in fragments. It was not until the 1920's that it was recognised, for example, that even particles of matter could behave like waves or particles. Later all natural forces began to be quantised, which was previously only possible with gravity.

The Strange Universe: Einstein, Quantum Physics and the ToE

myths and legends? Are you quite sure?

Unfortunately, I have to disappoint you. This scene is an authentic reflection of the strange phenomena that make 21st century quantum physics so uncanny. At least until someone observes the ghostly room. Strangely enough, what the theory (and every reasonable person) would expect. But nobody knows how the rooms know that you are being watched - and to top it all off, before you do it. The rooms do not behave correctly spontaneously as your children might do when you enter their rooms at bedtime. If your children want to sleep, read a comic or a booklet under the blanket. As soon as you open the door, the light is off immediately and your child is trying to sleep. As a conscientious parent you will of course think you know your children, and you let the secret reading pass as a "good sleep". After all, reading is not necessarily a rough verdict and some parents of adolescent children would wish to leave it at that. The spaces in quantum physics, or rather the mysterious phenomena that take place in them and do not even appear when you observe them. The phenomena do not stop as soon as you look, but already when you go to look. No one can tell today how the rooms can know you are looking into them before you actually do. Nor do we ask why a sofa is forcing itself into the kitchen and the television disappears without a trace. Not yet. The fact is, and you should already be aware of this: This quantum physical haunted castle is not a fairy tale. This is the bare reality. It embodies all the inexplicable phenomena that dominate the world of the small. Let us follow the trail of these phenomena and ask ourselves how these phenomena elude everyday perception. Einstein explained the photoelectric effect with the light quantum hypothesis. Since then it has been known that light is subject to a so-called wave-particle dualism. Light can behave as a particle or as a wave. Depending on the situation, one or other view provides the result that corresponds to the experimental results and hence to the correct result. In 1924, the French physicist de Broglie took one further decisive step. He postulated that even material particles could behave as waves in certain cases. Accordingly, an electron does not always appear as a punctiform particle but, depending on circumstances, which are yet to be explained, also appears undulating. This is the concept that a tiny globule should suddenly behave like a wave, which takes some getting used to. Or better: that every matter also has a wave characteristic. Or have you ever watched your car swinging happily back and forth in the garage, scratching half its body? Does the wife possibly accept no blame for the striking scratches on the bumper? Did you somehow wrong her? Is the guilty verdict something to with the quantum nightmare?

The problem is when the particles are to be considered as particles or waves. In fact, only one of the two phenomena ever occur simultaneously. At least, as soon as someone looks. What at first sounds matter-of-fact on the basis of the double-slit experiment of the German physicist Claus Jönsson which was not conducted until 1961, corresponding to a Pearl Harbour world view. This was a rather unexpected assault on the haven of physics. At the same time the philosophical flank suffered a veritable shipwreck. By the dawn of the 20th century a massive hail of bullets had already been fired into the conventional world view, even though it was thought that almost everything had been researched in physics. For this reason, the Munich professor Philipp von Jolly even advised Max Planck in 1874 to study physics: "Almost everything has been researched in this science, and there are only a few insignificant gaps to fill."

A symbolic quote for the biased views that were shared at that time in the rush to get to the next few decades jumping from one upheaval to another. Even though the questions that were posed then and now have been answered, we still know hardly anything. There is missing information in all respects if we are to close the few insignificant gaps that remain to be filled in physics. Mathematics. philosophy. physics. For the collapse of the classical world view in favour of a relativistic view should be only a scratch compared to the barrage which thundered out of the cannons of quantum physics. Jönsson's double-slit experiment permeated everything on an experimental basis that had previously been considered bullet-proof. By now quantum physics had become extremely disturbing. In 1961, Jönsson succeeded in examining the matter-wave hypothesis of de Broglie in a very complicated experiment. Complicated because it moved within very small orders of magnitude. It was to be the first double-slit experiment to be successfully conducted with classical particles instead of light (or electromagnetic waves). For this he devised a test structure such as that shown in Figure 7, a source (Q) from which electrons are shot at a double slit. There are two slits that enable the electrons to fly. Finally, the electrons that made it through the gap (and did not fly into the wall) hit a photo plate where it registered its impact.

The Jönsson double-slit experiment proved two phenomena that must have brought tears to the eyes of every scientific contemporary. Of joy or frustration. Not only because the subsequent observations were not only very strange but also, they were quite disturbing. First, the classical doctrine assumes that electrons are material particles and not waves. For if the building blocks of matter, i.e. electrons, protons and neutrons, were waves, the structures developed from them would behave accordingly. A billiard ball, for example, transfers its impact directly to another ball, clearly displaying a behaviour that would be assigned to a particle. A gun projectile drills into a wooden wall and leaves a clear bullet hole. This is also behaviour that you would expect from a particle, but certainly not a wave.

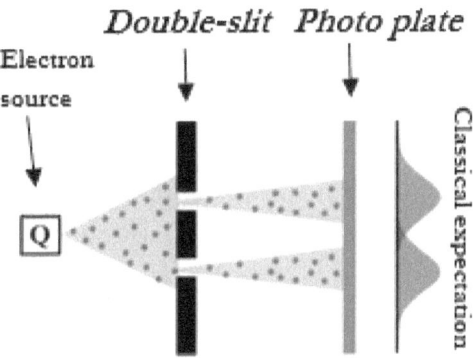

Figure 7   Double-slit experiment - expected result

Accordingly, the electrons can be thought of as bullets shot from the source Q onto the double-slit device. Some of the bullets catch one of the two slits and fly through the opening. Behind the double-slit device they collide with the photo plate. This photo plate registers the impact.

The two "peaks" on the far right in Figure 7 show the expected distribution of electron impacts. The electron impacts accumulate behind the two openings, but toward the sides fewer and fewer intuitively do so. These "peaks" make it easy to determine where the two slits are located, i.e. where most of the impacts are recorded. Toward the sides fewer and fewer electrons accumulate because there is a decreasing probability that an electron will get through the opening. No electrons at all reach the outer edge of the photo plate because to do so they would have to fly in a curve to the left or right after passing through the opening. If only one electron were to hit the photo plate it would be quite certain by the point of impact which slit the electron flew through. Except that the electron flies diagonally through the cracks and arrives right in the middle behind the two slits on the photo plate (where the two "peaks" overlap). The electron could then have passed through the left or the right slit.

So much for the classic expectation. So far so good.

The problem with this is that quantum physics is no everyday theory and is generally difficult to understand within the context of our intuition. The electrons seem behave randomly. The actual result of the experiment does not correspond in any way to the expectation described. The photo plate shows a pattern that is incompatible with everyday expectations. A pattern that once again proves how incomplete our knowledge of the world and its background is. In fact, the experiment formed an interference pattern (as shown in Figure 8). This means: Directly behind the slits (where most of the electrons would be suspected of being), the fewest electrons hit the photo plate. For this electrons accumulate in areas that are hardly or less likely to be accessible (for example, between the slits or next to the slits). This frequency distribution can actually occur only when waves are shot at the double-slit device.

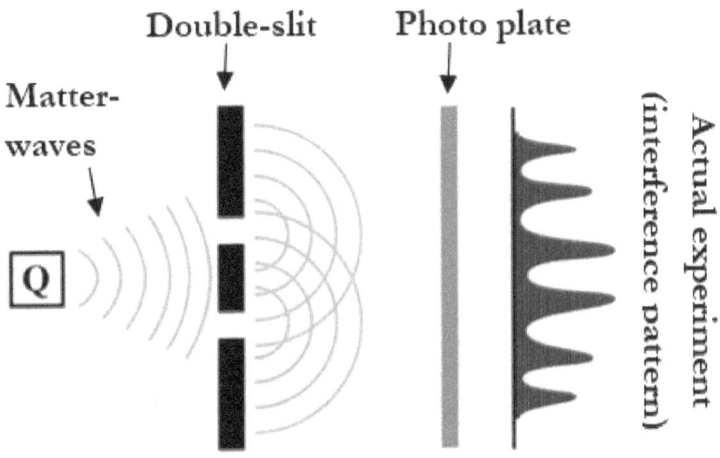

Figure 8 Double-slit experiment - actual result

For waves overlap each other, and this may strengthen them or destroy them.

This gives rise to the typical peaks and troughs on the photo plate, a so-called interference pattern. However, an electron is not a wave but a particle which may be conceived as a bullet. At least that's what you thought.

With the interference pattern, the experiment provided proof of the existence of the matter waves postulated by de Broglie. Even classical particles such as electrons or protons can under certain circumstances, for example in this experiment, assume a wave character. Electrons then behave as waves, although they are actually particles according to the classical view. Only in this way can the interference pattern be explained conclusively. Conclusive is however a gross exaggeration. The experiment has at least one guideline that causes physicists and philosophers alike to fall into mesmerising despair. This drawback, an anchor of physics stuck on the ocean floor, can be found in the answer to the following question: What happens when one of the two slits is closed?

What now appears on the photo plate is no longer an interference pattern but the classical mapping of the slits ("peaks"), as we would have already expected (in Figure 7) with two open slits. However, the real explosiveness only becomes apparent when we take the experiment one step further and ask ourselves: what happens when both slits are open and we try to determine the slit through which the electrons fly? Then, dear readers, the eerie fascination of quantum physics comes to full bloom. On the photographic plate, the classic image of the slits, that is the image we would have expected from our everyday conception, is seen again. Why does an attempt at interference appear once and for all in the same experiment as the classic image of the slits? And, by the way, why do the electrons behave as particles and as waves? And how do they and we know which behavioural pattern will be used and when?

The outcome of the experiment obviously depends on whether we try to determine the path of the particle or not. In other words, whether we observe the experiment or not.

We performed the first double-slit experiment without observing the electrons in flight. We only evaluated the impacts on the photo plate and found that an interference pattern was created. In the second double-slit experiment we tried to determine the slit through which the electrons flew. We noticed that a classic image of the slits was created on the photo plate, but no interference pattern.

The Strange Universe: Einstein, Quantum Physics and the ToE

Two identical experiments, but two contrary results.

This violates a fundamental principle of classical physics: The outcome of an experiment now depends significantly on measurements, although these measurements should not affect the experiment in and of itself, for the measurement itself has no interfering influence on the experiment.

In the classical view this phenomenon is inexplicable. Simply impossible. Imagine firing several shots from a pistol at a wall and this shows nothing but a concentration of most of the bullet holes in places that are barely reached with the pistol. Now attach a device to one of the two slits to see through which slit your shots actually hit the photo plate. And now. What happens now? Exactly. Now all your shots are concentrated more or less immediately behind the two slits - the experiment behaves as we would intuitively expect. But how is it possible that the outcome of the experiment depends on whether we measure it or not? And how is it possible that particles form an interference pattern and concentrate on the photo plate in places that they could hardly be reached in the classical view? In other words: how should the wave nature of material particles be understand? How can something be a wave that shapes consistent, solid matter like rocks, houses or cars?

Quantum physics gives us only one explanation:

As long as we do not try to measure the trajectory of the electron, the electron flies through both slits at the same time, taking all possible paths. It is almost everywhere. The electron behaves as a matter wave and can be superimposed on other electrons, which creates the interference pattern. The wave-particle duality therefore also applies to material particles. As soon as we observe the particle or try to determine with a detector which slit it flew through, the nightmare of the quantum world disappears. The particle now behaves as we know it from everyday life. Namely as particles. It travels along a single specific path, flies through only one of the two slits and leaves a single clearly defined impact point on the photo plate.

What is this supposed to mean? Has the experiment or particle developed a life of its own and does it behave differently according to whether a conscious observer is watching or not?

Researchers around the world argue about how exactly the experiment is to be understood. As long as the particle is not measured it remains in a quantum mechanical superposition state. Its probability wave indicates where it is located. Consequently it is in several places at the same time. Only when we intervene in the quantum system and want to measure the location of the particle does the probability wave collapse and the particle immediately decides on a permanent position.

The detector can have no influence on the particle in this case, at least no influence that would not violate the principle of causality and would require knowledge of future events. However, it is not the case that the detector would affect the path of the particle in some obscure physical way, or would explicitly destroy the other paths. Because, and this gets even weirder, the detector is behind the slits (!). The particle is observed only after the slits are crossed. Nevertheless, the particle already seems to know before reaching the slit that we want to measure its path. Because once we activate the detector it again follows the classical assumption and only flies through one or the other slit - along a single path, as we would expect from a proper particle (which can be thought of as a sphere). But if we switch off the detector, the particle follows the nightmare quantum world again and follows countless paths simultaneously. The particle is actually rather cheeky. It eludes direct evidence that it simultaneously flies through both slits and obeys mysterious quantum mechanical laws. Only indirectly, by the interference pattern on the photo plate, can its strange behaviour be confirmed. .

In any case, the detector does not measure the particle until it has flown through the slit. Consequently, the particle must already have known that we want to measure it before it has flown through the slits in the detector. Otherwise, the detector would register the particle at both slits simultaneously. The particle therefore has a kind of quantum-mechanical future vision. It already knows that it should behave like a classic particle, because it will be measured before the measurement has even taken place. The effect - its behaviour as a classical particle - converts the particle before it even knows the cause - the measurement. A very strange phenomenon.

Do particles have a sort of consciousness? Or how did this strange dependence on future events affect the outcome of the experiment? What is the explanation for this mysterious behaviour?

The double-slit experiment was to a certain extent the Michelson-Morley experiment of quantum physics, and poses numerous puzzles to this day.

3.3 Einstein, Schrödinger and the half-dead cat

Quantum physics lacks a mathematical basis rather than a

conclusive interpretation. Mathematical formulations and results are known, but nobody knows why it is the way it is. All physicists know the architecture of this haunted building, but nobody understands it.

The situation compares well with Newton's dilemma. Although he developed an equation with which the effect of gravity in everyday life could be described quite accurately, he had to admit that he did not know what was behind gravity as a basic force. The cause of the effect was completely beyond to him. Only Albert Einstein unveiled gravity in the general theory of relativity, attributing it to fundamental geometric principles and to the curvature and distortion of space-time. Even before nuclear fission was tested and the first aboveground atomic bombs were detonated in American deserts, the foundations of quantum mechanics had been created. Mathematically as well as physically. But we still lack a truly comprehensive interpretation that channels the quantum nightmare into a full explanation, eliminating all the question marks that haunt the microcosmos.

The surprising realisation that an unobserved particle follows multiple paths, that is, in many places at the same time, marks only the beginning of a long path of scientific reorientation. Indeed, quantum physics has no qualms about destroying just about all the fundamental principles of physics. The fact that a particle, as soon as we observe it, sheds its quantum chaotic character and meets our everyday expectation, marks the first step in a long journey of discovery. A journey of discovery which should inevitably lead to a superordinate theory inasmuch as our minds are able to follow the flow of truth, via all the waterfalls and whirlpools, to the source. Until then there is a long way to go. For as a fundamental theory, quantum physics would ultimately have to be applicable to all size scales (within Planck limits) or, in particular, provide an explanation as to why quantum phenomena are largely suppressed in everyday life. In the realm of the very small the world seems instantly to be disorganised chaos. The particles are a single wild bunch that runs counter to classical calculability. The well-behaved billiard balls become stubborn random elements that defy prediction and make the game with an uncertain outcome exciting again. However, as soon as we plunge ourselves into the city, sit in school, watch TV or cheer on our football team at the stadium, nature behaves as we are used to. In our everyday world the mysteries that shape the world of the small disappear.

In most cases at least.

The double-slit experiment is just one key to entering the strange house of quantum physics. But many roads lead to Rome. Another prime example, which continues the mysterious phenomena and ports to the middle of our macroscopic everyday life, is the "Schrödinger cat". The shadowy cat which brushes against the chair leg.

In the mid-1930's, quantum mechanics was in debate and the professional world was on the warpath. It was not necessarily about whether the theory was right in and of itself but rather whether it was complete. Einstein found one reason for this scepticism found in the uncertainty principle that the German physicist Werner Heisenberg set up in 1927. The Heisenberg uncertainty principle is a cornerstone of quantum mechanics and states that you can never accurately measure two complementary quantities at the same time. Therefore, you can never accurately determine the location and velocity of a particle at the same time, no matter how good your measuring instruments are. This statement is very important to an understanding of quantum mechanics. Its interpretation means, for example, that the motion of a particle cannot be predicted but can only be estimated with a certain probability. Since we cannot determine the momentum and location of a microscopic ball at the same time, it is in principle not possible to predict its trajectory precisely, but only with a certain probability. Nature has a veiling mechanism that prevents us from looking too deeply into it. The more precisely we try to determine the impulse, the vaguer the location becomes. And vice versa. This can be illustrated by taking a poker game. Imagine being invited to play poker with a colleague. Imagine that you are rather forgetful and can only memorise the cards you hold in your hands. You can only remember all the cards you normally see as long as you look at them. As long as you look at the table and see the cards lying there, you can make a decision based on your cards in your hands and the cards lying on the table. Once you take your eyes off the open cards, forget them and your decision will depend solely on the cards you hold in your hands. Of course, you could now be clever and improve your luck slightly by taking a look at the neighbour's cards – unobtrusively. Since you're pretty forgetful his cards can only influence your game decision as long as you look. So you either have the opportunity to look at the cards that are face up and include them in your decision or the cards of the opponent (let's say you play only in pairs). So you only ever know your own cards and the cards that are face up or those of the opponent. This conforms to the uncertainty principle of quantum mechanics: The closer you look at the cards that are exposed, the less you know about the opponent's cards and vice versa. You can only determine the cards you are not looking at with a certain probability. For example, if you hold two aces in your hands and two aces are on the table, your chances of your opponent holding an ace are zero, provided that he does not wear long sleeves. If your boss gives you a lecture and you hear a beautiful song ringing through the headset at the same time, you can influence the "understanding the lecture of the boss" by adjusting the volume. The louder the music, the lower the perception of the boss and the higher the attention to the musical pleasures. Turn off the music completely and the boss enjoys your undivided attention, but you do not know which song the radio is playing. Turn up the music loud enough, you hear every strum of a bass instrument precisely, but you do not understand a word of your boss's lecture. The two terms "understanding boss" and "listening to music" are complementary in this respect. The more attention you give to one, the less attention you give to the other. This is the so-called uncertainty principle of quantum mechanics. The more accurately we determine the impulse of a particle, the more inaccurate the statements we can make about its whereabouts. The uncertainty principle is inherent in Nature and is not due to the lack of precision of the measuring instruments. Are you also wondering whether you should take legal action against the police because in principle the radar devices are unable to measure your speed precisely?

In fact, a law enforcer in the particle world would not have it easy. Suppose Rolf, an electron in his prime, a friend and helper by profession, wants to set up a radar trap to ensure that his fellow species do not fly too fast through the atomic orbit. Rolf is spoiled for choice. He can have his speed camera made to a very high quality to achieve a precise determination of the speed at which an electron flies past him. The downside: If Rolf measures the speed of passing electron Jenny with a precision of 99.999 percent, Jenny's whereabouts become extremely blurred. Although Rolf knows pretty well how fast Jenny is travelling, he has no way of photographing her "electron number". Why? The uncertainty principle implies that the precise measurement of the property impulse (or in this example the velocity) makes the knowledge of the location very vague. Rolf knows how fast Jenny is flying, but has no idea where Jenny is when the speed is measured. Because Jenny obeys the quantum nightmare; she does not necessarily fly straight on, as you're probably most likely to do when driving your car (more or less gesticulating wildly) when you've been bothering your friendly PC for attention.

One reason for this blurring phenomenon is that in principle no measurement is possible without influencing the quantum mechanical system. To measure the impulse of a particle, you can fly it through a field or a light barrier. As a result, the particle is inevitably influenced and deflected from its trajectory. This is tantamount to the attempt by the police to determine the speed by placing a wall on the street and analysing the skid marks. The more precise we want to measure our impulse, the more intense the field or the light barrier and, consequently, the interventions in the system have to be. So although the impulse associated with a strong influence on the particle is known well enough, we no longer know where the particle is located. If we could determine the impulse with infinite precision, the location of the pulse would be correspondingly infinitely blurred and it we could glean nothing more about its whereabouts. Another cause of the uncertainty principle is the wave-particle dualism, the realisation that every particle also has a wave nature. The greater the maximum deflection of this matter wave, the greater the possible area in which the particle can be located.

The Heisenberg uncertainty principle introduces probability and thus a first approach of chance as a fundamental principle of quantum physics. It therefore destroys the familiar concept of a predictable world. In classical physics, the development of each self-contained system could in principle be predicted for all ages. In quantum physics, however, the outcome of even the smallest experiment is always associated with indeterminacy. It is therefore impossible to accurately predict the temporal evolution of a quantum system. Forecasting is possible only on the basis of probability considerations, just as we cannot know the outcome of a poker game in advance with absolute certainty without giving luck a helping hand. However, we can determine the probability with which we can win the game based on the cards we know.

For the time being, the question remains whether quantum mechanical systems are actually controlled by chance or whether their behaviour is due to our ignorance. There may be hidden variables that are not or not yet accessible to us at all. This could, for example, include additional dimensions from which the behaviour of the particles could be deduced. Without the four-dimensional space-time of the theory of relativity, numerous phenomena in high-energy physics, for example in particle accelerator experiments, could not be explained either.

The randomness of quantum mechanical processes has far-reaching consequences. If you are sitting on a comfortable couch reading this book, you probably feel relatively safe about the durability of your cushion. You would probably never have come up with the idea of questioning the goodwill of your sofa. This is a major fallacy. At least in the world of quanta. Your couch is only stable with a certain probability. Of course, it is very unlikely that your familiar seat will spontaneously disappear or assume an extremely obscure shape. But it is physically possible. Remember that all matter is composed of atoms and these atoms are made up of elementary particles such as electrons, neutrons and protons. These particles are only with a certain probability constantly underneath your posterior. But no one can accurately predict where these quanta will be in the next second, minute, or leap year. We are used to a certain consistency in everyday life and would probably be very insecure if a chair in the living room starts to play tricks or the TV were to run away (which one could understand when you look at certain programmes). But these are all phenomena that are at least conceivable in terms of quantum mechanics, even though their probability is very small. The consistency of everyday life is not a law of Nature or a fundamental principle. The couch does not sit under you because it cannot do otherwise but because it is very unlikely that all the particles will hit a very unlikely course at the same time. The fact that quantum phenomena have not yet found their way into our everyday lives is sometimes due to the fact that everyday objects consist of billions of particles. The chance that a high proportion of the particles at the same time will take a very unlikely path, turning your couch into a chair or simply marching into the kitchen, is very slight. Once again, I would like to make clear that the possibility, although almost infinitely small, still exists in principle. Just as most people never win the lottery and crash on the plane at the same time, most people will never see their sofa marching into the kitchen.

With the advent of probability and blur, the world becomes unpredictable to us. Not because our knowledge is too limited or our technologies too immature, but because it is a fundamental principle. Even with astronomical computers we could not calculate future development precisely but only estimate it with a certain probability. The more processes interact and the longer the period considered, the less accurate will be the forecasts.

The Strange Universe: Einstein, Quantum Physics and the ToE

Some of Einstein's contemporaries were unenthusiastic about the randomness of quantum physics. They did not attribute this to a fundamental principle of Nature but only to an incomplete theory. Finally, before Newton developed the gravitational equation, nobody could accurately predict the fall of an apple ("it is falling on the ground" is not "precise"). Einstein also disliked the quantum nightmare and he clarified his point of view with the much quoted statement: "The old man (God) does not throw dice". The Austrian physicist Schrödinger, meanwhile, set out to show the experts the incompleteness of quantum physics in a thought experiment. A cat and an unstable nucleus are in a closed space, such as an airtight box, as shown in Figure 9. There is no way to observe the events in the box without opening the box. If the atomic nucleus disintegrates, a mechanism is triggered that releases a poison gas, causing the cat to die. The probability that the atomic nucleus will disintegrate within an hour is 50 percent and 50 percent that it does not decay. Without opening the box, nobody can tell if the atomic nucleus has already decayed or not. Therefore, the nucleus is in a quantum mechanical superposition state[41]. It collapsed and did not collapse at the same time. As long as we leave the experimental setup to ourselves and do not observe, the possible states are superimposed. Already in the double-slit experiment we have seen that in quantum physics, states only assume a certain value when we measure them or destroy their isolation from the outside world.

In the example with the Schrödinger cat, the superimposition is destroyed and thus the familiar shape of physics materialises as

---

[41] The state overlap is also referred to as superposition in quantum physics. Here a particle finds itself in two states at the same time, for example destroyed and not destroyed. As soon as the particle interacts with the surrounding area, the superposition is destroyed and the particle decides the state it is to assume.

soon as we open the box. At this moment, the atomic nucleus spontaneously decides on a state (decay or not decay), which also determines the fate of the cat. Of course, the system cannot be outsmarted by, for example, mounting a radio-controlled camera inside the box. As soon as the camera sends its first signals the atomic nucleus jumps to a clear state.

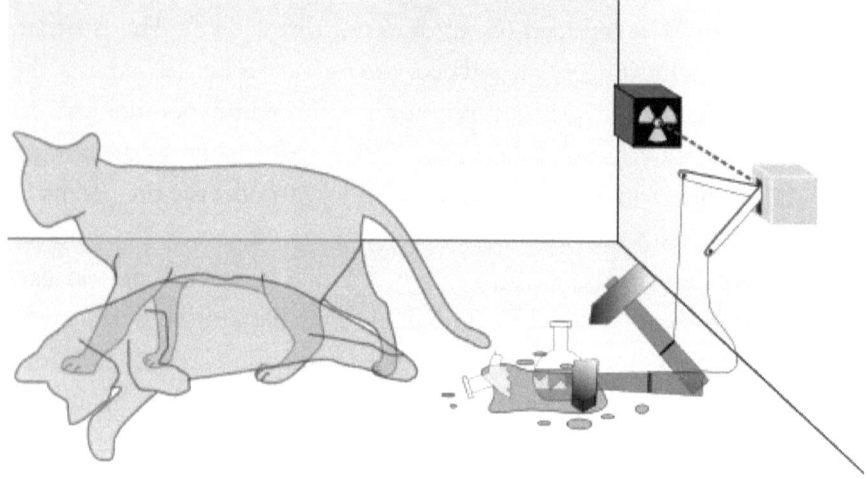

Figure 9   The Schrödinger cat

Schrödinger intended to use his thought experiment to point out the strange paradox that arises when quantum physics is completely valid, even in macroscopic spheres, in our everyday life. Because not only is the nucleus, but also the cat, is then in a superposition state. As long as the box is closed, the cat is neither dead nor alive, but half-dead (or half-legged). At least if the principles of quantum physics are uncompromisingly transferred to macroscopic dimensions. Such a superposition state is incompatible with our everyday experience. Nothing can be dead and alive at the same time. One excludes the other. Schrödinger showed in this approach that the quantum theories are still incomplete. There is no conclusive explanation as to why the quantum nightmare from the macroscopic world holds back gracefully.

In fact, over the years, different interpretive approaches have emerged to tackle this problem of correspondence. Otherwise, quantum mechanics would have been doomed to failure. We recall the principle of correspondence: Every higher-level theory must at least include all proved subordinate theories as a limiting case. This also means that in everyday life quantum mechanical phenomena are suppressed because there is (apparent) order before quantum mechanical chaos.

The Austrian physicist succeeded in proving an incomplete aspect of quantum physics with the Schrödinger cat. The crucial question is to what extent the cat is subject to the state superposition and how the paradox of the half-dead cat can be understood.

3.4 The Copenhagen Interpretation

Quantum physics embodies principally two unsolved problems. First the quantisation of gravity, and hence the union of quantum theories with the general theory of relativity, and second the interpretation of the strange phenomena that have grown out of experimental and theoretical knowledge. So far no one has really understood how the quantum nightmare is to be understood and what lies behind it. Nevertheless, two interpretations have prevailed in recent decades with the Copenhagen Interpretation and the Multi-World Theory. However, these are only a modest beginning in the endeavour to understand the phenomena of the quantum nightmare.

The Copenhagen interpretation was developed in 1927 by Niels Bohr and Werner Heisenberg. It is an attempt to interpret the mathematical framework of quantum physics. One key message relates to the coincidences, that is, the aspect on which Einstein cut his teeth. According to the Copenhagen interpretation, the indeterminable character of quantum physics is inherent in nature and is not due to a lack of knowledge or unknown mechanisms. In principle, it is impossible to make reliable predictions about microscopic processes, no matter how hard we try and no matter how well we understand the world of quantum science. According to this interpretation, a particle is in several superimposed states or paths simultaneously. As soon as it is measured, its probability wave collapses and instantly jumps to a clearly defined state, a state know from our everyday experience. In the thought experiment with Schrödinger's cat, the probability wave of the atomic nucleus collapses as soon as a conscious observer opens the box. At this moment, the atomic nucleus decides on one of the possible, previously superimposed states (decay or not decay), whereby the state superposition remains closed to the observer. As long as the box remains closed, no reliable judgment can be made about the condition of the cat. This also applies to the particles that are fired at the slits in the double-slit experiment. As long as no one measures which of the two slits the particle flies through, the particle moves on several paths simultaneously. At the same time the different particle states are superimposed. Only when someone wants to determine slit through which the particle flies does its probability wave collapse instantly and decides on one of the two slits. The state of the particle is not clearly defined until someone wants to access the particle. Before that, it is simply everywhere. At least if you follow this most popular interpretation among physicists - and believe me, this interpretation is relatively modest in terms of explanatory imagination compared to other theories that attempt an explanation of the quantum nightmare.

According to such an alternative theory, state superposition phenomena in macroscopic objects disappear even with the smallest interaction of the system with the environment. A box would not be enough to isolate the cat from the environment, which is why the cat's wave function and hence the state superposition collapse from the beginning. This explains the Schrödinger cat and, more generally, the lack of superimposition phenomena in macroscopic objects. A cat that is alive and dead at the same time completely defies our imagination.

Incidentally, the Copenhagen interpretation does not even attempt to assign the mathematical construct - for example, the wave function, which describes the behaviour of a system in quantum mechanics - to a real correspondence. Rather, this interpretation simply considers these mathematical properties as necessary for deriving the phenomenon in and of itself.

Incidentally, the Copenhagen interpretation is not the only attempt to interpret quantum physics. Two other candidates are considered plausible. Such as the multi-world interpretation. This assumes that every possible outcome of a decision in parallel worlds is actually realised. Reality splits up into countless universes every moment, taking every possible path. The supposed state superpositions would therefore not be real state superpositions, but each state is realised in a parallel world. The nucleus decays in our world while remaining stable in a parallel world. As already stated, the likelihood of a couch marching into the kitchen or mutating into an armchair is extremely low, but still greater than zero. According to the Multi-World Interpretation, every unlikely event is realised in a parallel world. So there is a universe in which you can win any lottery draw and a universe where you hit every flash of lightning that strikes anywhere.

Most physicists cannot accept the idea that there are innumerable parallel but inaccessible universes. Why should Nature be so wasteful and produce every possible condition? What sense would decisions make if everything happens anyway?

For as long as there is, albeit small, the probability of an event, this event takes place in a parallel world. Consequently, people in a parallel world could become extremely old. Every second and hour the probability of dying increases. As long as it is never really zero and thus a small probability remains, the human being would in principle continue to live in a parallel world.

A third interpretation of quantum physics, which seems more realistic at first sight, traces the quantum nightmare to hidden variables. Hidden variables that are located in higher dimensions or hitherto misunderstood aspects of modern physics and ultimately attribute the eeriness of the quantum world to the ignorance of man. Although the Copenhagen Interpretation, as arguably the most popular interpretation, postulates probability and chance as the fundamental natural principle of quantum mechanical processes. Newton was also convinced that he would be able to reach any speed if he just moved faster and faster. It took three hundred years for Albert Einstein to limit the cosmic speed limit to the speed of light with his relativity theory. Who would have thought at the time of Newton that gravity is due to a geometric principle of a four-dimensional space-time.

Even the interpretation and thus the real understanding of

quantum physics will probably only emerge from a more comprehensive theory. In any case, we cannot yet explain quantum physics, even though we've built it into a solid, mathematical framework.

3.5 The quantum tunnelling effect

Einstein contributed significantly to the development and success of quantum physics. Nevertheless, he fought with all his might against the mysterious phenomena, the eerie figure that seemed to impose itself on the theory of Nature. Not only that, the microscopic world tended to show a randomly governed character. Moreover, the theory of relativity suddenly resembled the rolling movement of a drunk, as theories and observations predicted phenomena such as black holes, time travel with a handful of paradoxes and the chronic instability of the universe. When Einstein conceived the revolution of the order of physics in the Patent Office in Bern, he had probably imagined the matter a little differently. He was to find no peace in this respect throughout his life. Physics began to get spooky. This left Einstein with an arsenal of unerring ironic quotes, the only weapon in the fight against this mushrooming science.

The revolution eats its own children. The construct of quantum physics, which was essentially based on Einstein's architecture, could no longer be destroyed. For its mathematically very precisely formulated statements seemed and still seem to be confirmed experimentally very accurately. The double-slit experiment was carried out tens of times in tens of laboratories in tens of variations without finding any contradiction to the formalism of quantum physics. But the Schrödinger cat and the double-slit experiment are only the beginning of an impressive phenomenology that we still cannot explain. A phenomenology that questions our view of the world in philosophical and physical terms. A phenomenology on whose interpretation scientists on all continents are cutting their teeth. Despair prevails. Everyone is waiting for the Einstein of the 21st century, who is sitting quietly and amiably in his little room and one day in the not too distant future putting the theory of everything on the table, the formula that unravels the quantum nightmare.

Einstein quickly lost his initial enthusiasm for quantum physics when the phenomena became too strange for him. He was less than thrilled when the tunnel effect and the "ghostly" long-distance effect left the world of physics completely non-plussed. The tunnel effect basically means that you only have to open a front door with a certain probability in order to get outside. For with good timing you could just as well simply walk through the wall. An obstacle is not always an obstacle. In quantum physics, the state of a particle is described as a wave function, as we well know. This indicates the probability with which the particle is located at a certain point. In classical physics, it was assumed that its location could be accurately predicted if there were enough data and parameters relating to the environment. In fact, the uncertainty of the location is a fundamental principle of our universe and is not due to a lack of knowledge. A particle is always in a certain place with a certain probability. An apple cannot leave the fruit bowl in which it is located on its own. Its skin represents an insurmountable obstacle for it. To lift the apple from its skin you have to exert a certain force, for example, to remove the apple from its skin by hand. Classical physics applies to the spheres we find in our everyday life almost without restriction. In fact, an apple that suddenly jumps out of its skin would create a very spooky impression. But in the world of the small, things looks rather different yet again. There, an apple can completely overcome its skin as if by magic. A particle, on an analogy to the apple, can overcome barriers by simply tunnelling through the obstacle. It sort of flies through the wall. This effect is called "tunnel effect" in quantum physics. The reason for this peculiar behaviour: A probability wave describes where the particle can be located. This probability wave goes beyond the obstacle into a classical "forbidden zone". Thus, a probability greater than zero exists that the particle will appear in this forbidden zone. Thus the apple can suddenly be outside its skin, without an external force So it is conceivable that a person could walk through a wall. All the particles from which a human being is materialised would have to appear in the forbidden zone at the same time, and he or she would have left the house, at least temporarily, without having walked through the door. Of course, the probability that all particles behave in such an abnormal way at the same time is almost infinitely small. It just means that it is not impossible, but rather unrealistic. Based on the phenomenon of the multi-world theory, every possibility, even if its probability is still that low, is realised in a parallel world. Thus there would be a world in which Hans, probably to his own astonishment, left the house without ever opening the door. The world would seem rather strange to us if the phenomena of quantum physics were to go astray in our everyday lives.

The tunnel effect is a very strange phenomenon. Purely mathematically it can be reduced to the probability wave. But what really lies behind it, in fact no-one knows how the tunnel effect works in space-time or how a particle leaves the fruit bowl It would definitely be wrong to think that a particle passes through an obstacle because, due to its size, it fits through a suitable loophole. The relationship between obstacle and particle can actually be illustrated taking the example of the apple and the fruit bowl. Intuitively, it is unthinkable that the fruit would leave the bowl on its own without external intervention. Equally inexplicable is the fact that sometimes particles simply overcome supposed obstacles. To this extent we know the phenomenon and a mathematical description. However, we have not understood the profound principle behind the tunnelling effect or the other quantum mechanical anomalies.

Not infrequently, the supposed anomalies of quantum physics also imply an attack on the pillars of the theory of relativity, in particular on the insurmountability of the speed of light. Until a few years ago Einstein's dogma was considered to be firmly established. Theories about an earthly overcoming of the cosmic maximum speed were frowned upon. Not only because it makes time travel into the past possible but also because the theory of relativity has been confirmed very well experimentally.

In 1992, an experiment in Florence showed that, as expected, the speed of light was not exceeded, even in the tunnel effect. But the German physics professor Günter Nimtz was not content with that. He mobilised for battle. With his assistant Achim Enders, he set up a test rig to unravel the mystery of the tunnel effect. To do this they measured the speed of microwave radiation in a constricted tube (waveguide). The results were amazing. The microwaves tunnelled the bottleneck, which is not possible according to the classical view. This provided further proof of the tunnel effect. So far so good. But the experiment went a decisive step further. The tunnelling speed of the microwaves was infinitely large, so it took no time for the microwaves to pass the obstacle.

Not a second. Not a millisecond. No time at all. It was to prove later that this is the case regardless of the tunnel length. Consequently, the microwaves pass through a fictitious tunnel from the Earth to the Sun immediately and without any time delay, while light takes over eight minutes. The hangman's verdict on Einstein's speed of light dogma?

But that was not all by a long chalk. Something very strange happened during the experiment. It turned out that microwaves always wait a short, constant time before entering the bottleneck. Almost as if they were taking a break to plan their journey through the tunnel. Professor Nimtz now began to smell danger. What followed was one spectacular experiment after another. In 1994, for example, he broadcast Mozart's 40th Symphony in microwaves through the constricted section of the waveguide. Some of the light used propagated at 4.7 times the speed of light. Should these attempts go down in history as the first experiments that violated the postulate of Special Theory of Relativity? Had Nimtz found what had been sought in vain for decades? Had the first irrefutable contradiction to the theory of relativity been discovered?

The experiment has been reviewed by several research groups, including Berkeley University. There the suspicion was confirmed that a supraluminal velocity results when there is an obstacle, a barrier, that has to be overcome between the quantum and the detector. The phenomena are once again obvious. This raises the question of arriving at the correct interpretation. Will the tunnel effect be the "guillotine" of the theory of relativity? Is there a surmountable but forbidden barrier behind the speed of light? Similar to the speed limit on the highway? Or are the strange results of the tunnel experiments due to measurement errors?

The Strange Universe: Einstein, Quantum Physics and the ToE

It has now been agreed that the measurements are correct and that measurement errors are excluded. The quanta actually tunnel at supraluminal speed. At least it seems that way to an outsider. The key question, however, is how we are to understand the tunnelling effect. Some physicists, such as Raymond Chiao, who has carried out tunnelling experiments at Berkeley University, suspect a mechanism that has not yet been fully understood to be the cause of the supposed speed of light. Similarly to the eerie distance effect, which allows two entangled quanta to "communicate" over light years without loss of time, the tunnel effect is probably based on a quantum mechanical principle, a principle that it has not yet been possible to explain with our theories. The only point of agreement is that the signal speed of a particle can never exceed the speed of light. Nobody knows what exactly happens in the tunnel or in the constant time before entering the tunnel. Not even how the apparent speed of light comes about. Possible explanations would be a kind of cosmic background field, a hyperspace or an oversized channelling of the universe.

Whether or not the dogma of Einstein is now questionable is not really known. Some people are convinced that at least information such as predicted by Einstein's Theory of Relativity can only be transmitted at the speed of light. Previous experiments conducted by the research group led by Professor Chiao showed that a high proportion of the photons is lost during tunnelling at supraluminar speed. This raises the question of the extent to which information can be transferred at supraluminar speed and hence even speak of communication. Other physicists argue that the alleged speed of light is only a matter of defining speed. An artefact of theory to some extent and thus no contradiction to the theory of relativity. Just as two cars on the highway travelling in opposite directions move apart at around 300 kilometres per hour, even though each individual vehicle only runs at 150 kilometres per hour. However, the speed of light cannot be exceeded by two particles moving at 90 percent of the speed of light travelling in opposite directions. They do not distance themselves from each other at 180 percent of the speed of light, as might be suspected in the classical view. Due to the high speeds, a relativistic calculation is required where the speeds always remain below the speed of light.

There is still no evidence that the theory of relativity should be revised, for in the tunnels the theory of relativity cannot be applied because it is not a valid reference system. Rather, as Nobel Prize winner Richard Feynman postulated, virtual particles that materialize back to normal particles at the end of the transition are transmitted. Such virtual particles play an important role in the context of the quantum nightmare in a vacuum, seemingly matterless space.

Richard Feynman was one of the great physicists of the 20th century who made a significant contribution to quantum electrodynamics. Feynman was also the man who recognised the essence of quantum physics at an early stage and summarised it as follows: "Anyone who has understood quantum physics has not understood it." Or, to reconcile the theory of relativity with quantum theory, at least rhetorically: "There was a time when newspapers were saying that only twelve people understood the theory of relativity. I do not think there was ever such a time. On the other hand, I think it's safe to say nobody understands quantum mechanics. "

The tunnel effect should not be the last phenomenon on which physicists cut their teeth. At least as strange is the ghostly long-distance effect, as Einstein devalued the following phenomenon in order, unmistakably, to express his criticism of the eerie evolution of quantum physics once more.

3.6 The eerie long-distance effect

Although, for every quantum phenomenon there is an illustration in our everyday life that does not make the corresponding effect unrealistic[42], it ports correctly to our everyday world in quantum mechanics. The tunnel effect allows humans to run through walls. The state superposition (superposition theory) allows a cat to be dead and alive at the same time. Given the eerie long-distance effect, however, all analogies from our everyday world fail. There is only one word we can use to describe the effect of this phenomenon in any way accurately: teleporting.

Two entangled particles[43] are distant from each other. Particle alpha remains on Earth, whereas particle beta is transported to the next star. The particles may assume two possible states, let us say "up" and "down". As soon as one of the two particles is measured, the other entangled particle immediately rises to the opposite state.

As long as we just let alpha rest, the particle is in state superposition. It can be both "Up" and "Down". When we measure the state of the particle, its wave function collapses and it assumes one of the two possible states. At the same moment the

---

[42] In this context unrealistic means that a quantum phenomenon can take place as an everyday occurrence, but the likelihood of this is very slim (and hence unrealistic).

[43] Two particles are entangled when they have formed in the same place at the same time. These particles are bound together from then on, but the mechanism behind it is completely unknown.

other particle jumps to the opposite state. The problem is that the measured particle randomly chooses one of the two states. In any case, the other particle immediately assumes the opposite state. But how does it know to which state the measured particle has jumped? How does beta know, on the distant star, what state alpha has just assumed on Earth?

Some physicists suspect hidden variables in which the state distribution is fixed from the beginning. Thus, the choice of state would seem only random since the hidden variables escape our view and access. The theory of hidden variables is due to experimental findings of recent years, but on shaky ground. And even if this theory is correct it cannot really explain the eerie long-distance effect. For the strangest aspect of this is that both particles, no matter how far apart they are, immediately assume a particular state when a particle is measured. However, according to the theory of relativity, information may be transmitted at no more than the speed of light. Otherwise, the causality would be violated, that is, the information reached the receiver in the past and thus before any information has been sent at all. So how is it that beta on the star, which is a few light-years away, immediately learns of the measurement of alpha on Earth? Without any time delay?

Immediately a possible parallel to the tunnel effect becomes clear. As with the tunnel effect, the speed of light also comes into question here. There appears to be an immediate transfer of the measurement where the distance between the two entangled particles can be of any length. Two entangled particles are connected by an unknown mechanism. Changes of state are transmitted from one particle to another as if teleported. The theory of relativity follows the principle of locality, which means that each event can only locally influence another event. In other words, if an event A takes place on Earth and an event B on Alpha Centauri, and both events are only one light-year apart, the two events cannot be affected since light takes four years to reach Alpha Centauri from Earth.

In quantum physics, a particle is only found at a certain location with a certain probability. This probability is described by a wave function. An electron orbiting an atom on Earth has the greatest probability of being near the atomic nucleus. In fact, the wave function also exists a long distance away, for example on Alpha Centauri. Although the probability of electron residence is very small, it is not zero. Once the particle is measured, the wave function collapses and decides where to place the particle. The key question is whether this collapse of the wave function occurs immediately and without time delay, or whether it propagates only at the speed of light and takes four years to reach Alpha Centauri.

As some television shows teach us, using one or the other joker is a worthwhile idea if you cannot answer the million dollar question. The public joker does not really take us further because of the chronic absence of an audience when these lines are being read. The Fifty-Fifty Joker would halve the hotchpotch of answers, but unfortunately we have nothing that could be cut in half. But of course we could call a friend, which is the joker we play to get an answer to our question. We call Mr. and Mrs. Quantum Physics. Of course, quantum physics helps us and replies that the collapse takes place immediately and is thus transmitted faster than the light. In fact, even at an infinitely high speed, there is no time at all until the collapse of the alpha particle on Earth has been transferred to beta particles on Alpha Centauri. This answer is of course correct. We thank quantum physics and enjoy the blessing of the money that such quiz shows brings us. But stop. The moderator offers us a bonus question. Double or nothing. We think that sounds fair and and consider the all-important question: how does the particle alpha transmit the information of its collapse in real time to the distant particle beta?

Now it becomes clear why Albert Einstein called this phenomenon an eerie long-distance effect. Although we play all the remaining wild cards and look to the moderator for help, we cannot answer that question. The quiz show is unfortunately lost. In fact, you will get about one million Euros if you answer that question. Behind it lies one of the biggest riddles of quantum physics. A riddle whose solution the Nobel Prize Committee would be only too willing to reward with gold plate.

Although the collapse of the wave function appears to be instantaneously transmitted to the entire universe, it does not necessarily violate the light-speed dogma of relativity. Various experiments have shown that with the eerie long-distance effect no real information is transmitted, that is, it is not possible to instrumentalise the eerie long-distance effect as a teleporter. For according to today's interpretation of quantum physics, the particle jumps to a random state. In principle, it cannot be predicted which particle will assume which state. Therefore, no data or even people can be teleported by means of remote action. At least not according to the current state of knowledge.

The Copenhagen interpretation, the most popular interpretation in quantum physics, has no explanation for or interpretation of the entanglement phenomenon. The multi-world theory is almost as perplexing and blames this on an unknown mechanism of a parallel world. However, there are other theories that have so far been ignored by experts. Thus a theory assumes a background field, which connects the particles outside the space-time and thereby allows the transmission of the measurement event in zero time. Possibly, this theory also helps explain the tunnel effect whereby the particles are transported as virtual particles over the background field and materialised after the bottleneck to an ordinary particle. Perhaps only a unifying theory of all the fundamental forces and elementary particles will explain this phenomenon. A hot candidate for such a theory of everything is string theory, but it is still insufficiently developed and understood to fully explain the enigmatic quantum phenomena. Sometimes all that is clear it that Feynman appears to be right. Quantum physics is a discipline of science that nobody really understands and it should stay that way for a while.

3.7 The quantum nightmare in a vacuum

In 1931, a 75-year-old electrical engineer drove a Pierce Arrow through Buffalo for a week, reaching top speeds of around 150 km/h. This was reported by various American magazines because the luxury car is not just any car. It has no engine, at least no petrol engine.

We are talking about Nikola Tesla, one of the most important inventors of the 20th century. He pioneered alternating current and researched the possibility of transmitting energy wirelessly. In the US alone he filed 112 patents, including patent number 685,958.

Tesla claimed that in our world there is an inexhaustible, freely accessible and omnipresent source of energy that can easily be drawn from the space around us. To use this energy, he constructed a round electric motor with a diameter of about one and a half metres, which was intended to replace the petrol engine in his luxury car. This electric motor derived its energy from a radiation receiver or gravity field generator located in front of the dashboard. It consisted among other things of twelve tubes and two strong rods which had to be pushed in to start the energy intake. Only Tesla knew the exact construction of this radiation receiver. At the beginning of the 20th century, the same invention was often made by different people almost simultaneously. Tesla wanted to protect his ideas, which is why he often omitted a crucial part in his design sketches. No one could replicate his ground breaking machines without knowing his plan.

Today we are in the 21st century. The world is struggling with a massive energy crisis, triggered by the shortage and drying up of the most important raw materials. The age of the oil company is drawing to an end. Technology and the economy are on the verge of sustainable change. But you will still hear nothing about Tesla's free energy. Nobody talks about using the natural radiation or zero-point energy with which the electrical engineer ran his car. Nobody wants to tap the perpetual motion of nature. Why not? The reasons are obvious.

First, the energy industry abhors the idea of a free source of energy that can be made accessible to anyone with relatively primitive equipment. Energy is a huge business. Trillions of dollars are earned from it every year. And this is a strongly growing trend in view of the economic growth of populous countries such as India and China and the exploding prices of raw materials. Of course, the economic argument is not enough to explain the renunciation of free energy. The question arises whether such an energy source exists at all and whether the radiation receiver, which is said to have powered the car, has ever existed. Although related patents that fundamentally protect such technology are registered in the USA, there is as yet no trace of the equipment that Tesla is said to have used in his car in 1931. Has the device been disposed of to protect the powerful energy lobby? Or has the device ever existed? Is there any free energy at all?

Fact: Tesla filed US Patent No. 685,958 for the use of radiant energy on November 5, 1901. This and other strange-looking patents actually exist. Another fact: Tesla was a very talented inventor who entered fields hardly anyone else would venture into. He was convinced, for example, that sooner or later machines would no longer need direct access to electricity because the energy would be transmitted wirelessly across the atmosphere. And indeed, reputable universities and research labs are currently trying to recharge batteries from portable computers without cables, but based on transparent electromagnetic principles. Free energy - all just hocus-pocus or maybe there is a grain of truth behind it?

With regard to the theory of relativity in particular we have repeatedly emphasised that the speed of light in a vacuum represents the maximum permissible speed of our universe. But what is vacuum actually? In school textbooks[44] vacuum is conceived as space devoid of matter, such as that which can be found in distant regions of the universe or produced artificially in laboratories. In fact, vacuum is not completely empty. Ideally there are no particles and fields of the standard model. There are no electrons, protons or electromagnetic fields in a vacuum. However, vacuum is not empty either. For the uncertainty principle of quantum physics prohibits an absolutely empty space and hence a

---

[44]   ... and to simplify certain circumstances, for example, in previous chapters of this book.

"sharply determinable" energy state. We recall that with a particle, location and momentum cannot be measured exactly at the same time. Vacuum is not therefore dominated by gaping emptiness, but is subject to the zero point fluctuation. Is this zero point fluctuation the free energy that Tesla wanted to make available? Does this quantum nightmare exist in a vacuum at all, or does it exist only in theory?

The Dutch physicist Hendrik Casimir postulated in 1948 a rather strange phenomenon, which went down in history as the "Casimir Effect". In a vacuum, two plates that are uncharged and parallel to each other were to be compressed by an invisible force. This force was not directly due to one of the four fundamental forces of physics, but much more to the inherent life of the vacuum. Or rather: to the zero point fluctuation. This creates virtual particle-antiparticle pairs and destroys itself instantly. Actually, it is not possible, at least so classical physics states, for energy to come from nothing. But as the particle pairs in turn instantly annihilate each other, the law of conservation of energy is not violated. The Casimir effect can be explained by this very zero point fluctuation. For certain states of the virtual particles allowed outside the plates are prohibited between the plates. As more virtual particles are allowed to hit the plates outside the plates, a pressure difference arises which is noticeable in the Casimir effect and compresses the plates. The smaller the distance between the plates the fewer virtual particles are allowed in between. As a result, the pressure increases the closer the plates approach each other (which is quite problematic in nanotechnology applications). Marcus Spaarnay experimentally proved the Casimir effect in 1958. The quantum nightmare actually exists in a vacuum. But is there any proof of free energy? In fact, the energy that seems to be in space in every vacuum and even at absolute zero is hardly usable as an energy source. This is because the virtual particles are beyond our reach. Should it nevertheless be possible to draw antiparticles from the zero point fluctuation and allow them to react with matter to provide usable heat radiation, the law of energy conservation would definitely be violated. In this case, the question arises as to where these mysterious virtual particles come from. If a conclusive answer is found, the energy conservation law could still be saved. In the bathtub we also have the feeling that the water is lost when we pull the plug out and drain everything in a vortex. If we look at the entire water cycle or the ecosystem, of course, no water is lost. At most it changes its state of aggregation (freezes or becomes gaseous) or reacts with another substance. Mass and energy remain in the global system but are preserved. Perhaps a similar principle applies to the virtual particles as well. They may only appear to us virtually, invisible to us, because they interact from a higher dimension or from hyperspace surrounding space-time. Free energy is conceivable, at least theoretically. The Casimir effect shows that the zero-point fluctuation compresses two plates and thus causes a force to act. Maybe some day it will be possible to use this force in nano applications, which would make it possible to tap vacuum energy, an energy that is available virtually everywhere and is almost unlimited.

In this context, of course, it would also be interesting to discuss possible alternatives. Maybe Tesla did not focus on quantum mechanical zero-point energy but on another freely accessible form of energy that fills the universe. Candidates here are neutrinos, dark energy or cosmic rays. Every second your hand is crossed by over a trillion neutrinos emitted by the sun. The neutrinos would be an almost inexhaustible source of energy. However, it is not yet known whether neutrinos have a mass and to capture even half of the neutrinos one would need a lead wall with the astronomical thickness of one light year or around 10^{16} metres (approx. 100 million times the distance between the Earth and the Sun). Neutrinos are therefore rather unsuitable as energy suppliers. Maybe Tesla might have meant the dark energy or cosmic rays that gave him the energy for his car. Astronomers believe that our universe is largely filled with invisible dark energy and dark matter. However, whether these can be effectively used on Earth as energy remains an open question. as is the question of whether cosmic radiation is strong enough to operate a vehicle. Incidentally, Tesla himself believed that decades or centuries would pass before mankind really made use of the radiant energy, be it for technical reasons or because of economic conflicts of interest.

3.8 Antimatter

Say we are writing in the year 1928.

In the US, the global economic crisis is waiting to sink the economic optimism on Black Friday. In Germany, the way out of a painful past is to enter an even more painful future. Between Buenos Aires and Melbourne a ship with 80 crew members disappears without a trace. In the UK, women are given the right to vote.

Meanwhile the world is changing in the quiet chamber of an adolescent genius. He is a young British scientist with Swiss roots immersing himself in the foundations of modern physics. He is convinced that the two pillars, quantum and relativity, are based on the same principle, the principle that underlies the world plan.

The Strange Universe: Einstein, Quantum Physics and the ToE

His name is Paul Adrien Maurice Dirac. He was one of many who were to take on this difficult task, and one of the few who succeeded in bringing light into the darkness. In his efforts to combine quantum physics with the Theory of Relativity, Dirac gained an extremely significant insight. Possibly one of the greatest secrets of nature, which raises questions going right back to the Big Bang: We are not alone. Dirac was looking for a neat solution to a problem that had arisen two years earlier. At that time, the Austrian physicist Schrödinger had published his "Schrödinger equation" and thus delivered the fundamental equation of quantum physics. It described how particles, also called quanta, behave. Yet there was a snag with the Schrödinger equation which deeply concerned Dirac. Since Einstein's theory of relativity it was normal to view three-dimensional space and time (the fourth dimension) as on a par with each other.

Every movement in the space known to us causes a movement in time. As you move from your home to the bus stop you move through space and time. In Schrödinger's equation, however, space occurred as a power of four, whereas time occurred only as a power of one. No trace of equality here. Or would you be thrilled if you, as an insurance salesman, earned two thousand Euros a month and your neighbour were to earn four million Euros for the same work?

Probably not.

That's what Dirac thought. The Schrödinger equation was a physical discrimination of time, so to speak. Dirac wanted to eliminate this situation. He tried to link Einstein to Schrödinger, or rather create a bridge between the two cornerstones of modern physics. And he was to succeed in bring some light into the darkness. He "toyed" a bit with the Schrödinger equation, as he later said, by combining the fundamental equation of quantum mechanics with the theory of special relativity. The square in the formula disappeared. Time and space were henceforth equal. Even for the neighbours only two thousand Euros were now paid for the work. As a result the equation was much more complicated than it had been before. But that was not the only tribute that Dirac had to pay. The "bill" should be much higher, because in his scientific zeal Dirac had forged a double-edged sword. On the one hand, the Dirac equation was compatible with the special theory of relativity and for the first time described another important property of elementary particles, the so-called spin. The spin is a kind of rotational movement, but cannot be explained classically (that is, with the physics we know from everyday life) That means: for the spin, there is no realistic representation in the "normal" world. On the other hand, the Dirac equation now produced very strange results, namely negative energy states. Thus a particle can have less energy than no energy at all. Admittedly a strange idea.

Dirac was not deterred by this. He was convinced that every mathematical result in nature makes sense. He trusted in the power of mathematics and accepted the negative result as a negative state of energy. Initially without knowing what was behind it. Should there be particles that move more slowly than not at all? Should there be particles that have less energy than no energy at all? Should there be particles with less matter than no matter at all? If Dirac was right and his equation was right, there had to be something that weighed less than nothing. A mass with a negative weight. What on earth was that supposed to mean? What can weigh less than nothing? The physicists put their hands in their heads. Is the world stalking a strange form of matter that we have not encountered before? Or is completely unknown to us from everyday life? Dirac was convinced. Trivially the negative result differs from the positive result only by the sign. The difference between a debit and a credit balance in the bank account is mathematically only one sign. That's why it stands for nothing but its opposite. The opposite of debit is credit. The negative result of the Dirac equation is nothing more than the counterpart of matter. Antimatter. Our world is not alone. And suddenly both results made sense.

Science did not really know what to think of this. There came one man, Dirac, who had previously been considered quite serious, who postulated, on the basis of a sign, the existence of a strange antimatter that hardly anyone could imagine - and not even want. But his extremely strange favour did not find favour everywhere. It did not help that Dirac grew up in the idyllic Valais peaks or had recently studied at the prestigious Cambridge University. After all, the existence of this alien matter was based only merely on "experimenting with equations"[45] and therefore it was quite hypothetical. The world of physics may still be reeling from the blow suffered by scientific revolution that had occurred at the turn of the century. The prediction of antimatter did not have a particularly healing effect on old wounds, and hardly anyone dared to imagine how far-reaching the experimental proof of antimatter would affect the world view.

The prediction of antimatter was definitely daring. A defensive

---

[45] Antimatter, mirror worlds, particles flying back into the past (tachyones) – the assumption of numerous rare phenomena in physics is attributable to the interpretation of different mathematical solutions to the same equation. However, until experimental evidence is produced no one can say with certainty whether these solutions have an equivalent in nature.

move against the rising current. But only four years later, the American physicist Carl Anderson provided the decisive proof. In an experiment he discovered that cosmic radiation releases foreign particles, called positrons, when it permeates matter. Positronsare nothing more than antielectrons, so to speak, the electrons of antimatter. This proved the existence of antimatter. Antimatter actually exists. Anderson received the Nobel Prize in 1936 for this sensational discovery. Dirac also reaped the fruits of his work and received the coveted award later that year. Dirac had searched for a relativistic formula and found antimatter. One of the greatest discoveries of the 20th century had become a fact. A discovery that has, however, only in recent years met attracted interest outside the professional world.

One key question remains: what is antimatter?

Let's take a look at nature, at the fascinating world we live in, and at its engine, its drive. Things in nature do not just go their own way, as you may occasionally go to the nearest pub for a beer after work. The movement of nature in all its diversity is based on the presence of imbalances and a natural urge for balance. We do not need to postulate any physical theories to confirm this statement. A glimpse of our own lives is enough to at least realise that previous assumption cannot be completely wrong. So we eat when we are hungry. Or balance on the bike so that we do not fall over. Or we go to bed when we are tired - or when your half asked to do so. In order for an imbalance to occur in our world in the first place, each physical existence requires a suitable counterpart. For example, protons (positively charged) and electrons (negatively charged) equalise each other. Or the earth balances the gravitational pull of the sun with (at least approximately) elliptical orbital motion, allowing it to maintain a more or less constant trajectory[46]. Or warm and cold ocean currents provide a reasonably even climate. In recent years, the idea of symmetry has prevailed in physics, whereby there is a counterpart to everything. Cold and warm. Plus and minus. Attraction and repulsion. In this respect, antimatter is nothing but the counterpart of matter. The elementary particles of antimatter are charged in exactly the opposite direction. Antielectrons (positrons) have a positive charge and antiprotons a negative charge. The other properties of the particles and their antiparticles do not differ. Antimatter, contrary to popular literature, is not an unworldly, inaccessible substance, but the opposite of matter in terms of particle charges. One of the biggest unresolved puzzles in the universe is the question of why the Earth, the Solar System, the galaxies and possibly the entire Universe are made of matter, not antimatter. Why have no antimatter suns or antimatter galaxies been discovered so far?

A valid question. If you can find a conclusive answer to this, you will win the Nobel Prize (which is worth 10 million Swedish kronor), with a personal congratulation to yours truly. When we look out of the window we usually see trees, roads, cars and maybe a lake or even the sea. No matter where we look. Everything we see is matter. When we peer into the depths of space with a telescope we see planets, stars and maybe even galaxies. No matter where we look. Everything we see is matter. We can search where we want. We will not find even a lump of antimatter anywhere. The antimatter plays a cosmic game of hide and seek. Where is there a balance, if there are more stars in the universe than grains of sand at the seaside, but not a single lump of antimatter?

---

46   Actually the Earth moves 10 cm further away from the Sun every year.

Well, we can certainly explain why there is no antimatter in the universe, or at least we have not discovered any antimatter. Antimatter and matter have opposite particle charges. The electron of antimatter (positron) is positively charged, the electron of matter negatively charged. As we know, positive and negative charges attract each other. When a lump of antimatter appears anywhere on Earth, it immediately reacts with matter, releasing vast amounts of energy, so-called annihilating radiation. That there is no consistent antimatter in our world may be due to an excess of matter. At the slightest contact, antimatter and matter burst into pure energy. Although antimatter can occur in the short-term in the high-altitude beam experiment or in particle accelerators, it is immediately destroyed in its reaction with matter.

Why does nature keep antimatter under wraps? Why are there countless planets and stars, but apparently no consistent antimatter in our universe? Does nature have a preference over matter? Is this all just a question of the horizon or rather of a more profound principle? Is there a good reason why we live in a material world? What is the cause of the symmetry break between matter and antimatter?

The existence of antimatter galaxies or complete universes outside our spectrum of observation cannot be ruled out. In another dimension, in a mirror world. Or billions of light years away from the Milky Way. We know today that the universe is expanding, maybe because somewhere in the depths of the universe antimatter galaxies exist and slowly link up to the matter galaxies. If, just a few thousand years ago, 99 percent of the universe known to us collided with a similar number of antimatter galaxies, we would only be able to guess at this cosmic spectacle in a few thousand years. At the earliest.

The Strange Universe: Einstein, Quantum Physics and the ToE

All the information we obtain from the depths of the universe is based on radiations that were emitted an incredibly time ago but are only now arriving on Earth. The light and any information in the universe can travel at a maximum of the speed of light. When a flash of light is fired at Earth on Alpha Centauri, the nearest star system, it will not reach us for about four years. When scientists talk about the fact that no antimatter galaxies can exist because no annihilation radiation has been observed so far, this means that there was no collision of antimatter and matter galaxies until a few million or billion years ago. What exactly happens at that moment in the depths of the universe cannot be determined until the witnesses of these events reach us. It takes about 2.5 million years for the light of the nearest galaxy (Andromeda nebula) to reach us. Ergo it would be possible for the Andromeda Nebula to crash into a black hole two million years ago - and we would not know this for another 500,000 years. The view into the evening sky is nothing but a look into the past. But no one knows what is really happening in the depths of the universe at this moment.

The current theory of everything candidates assume at least ten dimensions in our universe. It is quite possible that antimatter also found its place in the cosmos somewhere in a dimension that is inaccessible to us. A dimension in which matter is alien and rare. In a mirror world or inaccessible part of the universe. It is possible that all antimatter was destroyed at the birth of the universe and hence completely burned into energy. This may explain why, even in the most remote areas of the universe, it never gets colder than about three degrees above absolute zero (- 273.15 ° C). On the other hand, it follows from this theory that Nature must have a preference for matter, otherwise the universe would not have arisen.

Some researchers argue that the mysterious gamma-ray burst, huge bursts of energy in the universe, is due to the collision of matter with antimatter residues. However, a gigantic $1.1 * 10^{28}$ kilograms of antimatter would be necessary to produce, for example, the Gamma ray burst GRB-990123 that space telescopes recorded in January 1999. This is equivalent to the weight of about 1674 earth masses, a small antimatter sun. On the other hand there is no conclusive explanation for this tremendous phenomenon.

You may be shaking your head and wondering how science can claim to have discovered antimatter, if it exists at most beyond our horizons - in speculative mirror worlds or inaccessible areas of the universe - if at all. Well, you're right, of course. However, so far we have only talked about the occurrence of constant antimatter in Nature. In Nature, antimatter may have only existed in the past, in hitherto inaccessible dimensions, or billions of light years from Earth[47]. Say: Beyond our horizons. However, a research group headed by Professor Walter Oelert succeeded in 1995 in artificially producing antihydrogen in the particle accelerator at CERN in Geneva. The first experimental evidence of a so-called CP violation emerged. This effect simply means that you do not always look the same as your own mirror image. Or, to put it in the world of physics, Nature may have a preference for matter or antimatter. The question is how and why this imbalance can arise. Why is the universe made of matter and not antimatter?

The Chinese physicists Tsung Dao Lee and Chen Ning Yang postulated as early as 1956 that the reflection of a particle and its

---

[47]   Apart from unstable antimatter of course, as was demonstrated in cosmic radiation. In this case, however, antimatter and matter dematerialise immediately into pure energy.

antiparticle do not always have to be identical. One year later, physicist Wu succeeded in experimentally proving so-called parity violations that occur in the weak interaction. This therefore destroyed the prevailing idea that all laws of Nature are symmetrical, that is, that they can be mirrored without the original being different from the mirror image. CP violations (in some form) were probably essential in order to create the material world in which we live. Without this enigmatic phenomenon, whose cause and expression are largely unknown, the universe would have been destroyed immediately after a possible Big Bang.

The lifespan of the antihydrogen atoms artificially generated in 1995 was of very short duration, so they could not be investigated further. Only seven years later, researchers were able to produce about 50,000 antihydrogen atoms. Still a relatively small number considering that a single drop of water consists of billions of atoms. In this experiment in 2002, the antimatter was to be stored and cooled to near zero point temperature. It was hoped to find out whether antimatter complied with the "CPT Theorem", a fundamental law of physics. The "CPT Theorem" states that if we look at a process in mirror image, time-reversed, and also reverse matter and antimatter, the physical laws of Nature continue to apply to this process and thus such an operation is possible. The key question was figured out in terms of whether an anti-car drives the same as a car and an anti-meteorite really behaves just like a meteorite. Are antimatter and matter really different only in terms of the charge of their elementary particles? That is exactly what the statement of the "CPT Theorem" would state.

The experiment did not get that far, however To date, no researcher has been able to store antimatter, that is, to keep it longer than a few fractions of a second. As soon as antimatter and matter come into contact, the oppositely charged particles dissolve into pure energy in the most violent reaction. This process is called "annihilation". All the colliding masses are converted to an unimaginable amount of energy according to the famous Einstein equation $E = mc^2$. By comparison, the nuclear bombs dropped in Japan in August 1945 converted a relatively small part of the explosive material (uranium and plutonium) to energy. Within a radius of two kilometres everything was razed to the ground. If the Americans had equipped their bombs with antimatter, the detonation would have left total devastation and destruction as far as South Africa. At least.

Incidentally, it is a widespread misconception that the formula $e = mc^2$ would have led to the construction of the atomic bomb. For the formula only indicates how much energy a particular mass contains or how much mass could be obtained from a given amount of energy. We therefore know that the mass of our car has more energy than any atomic bomb. However, the formula gives us no clues as to how we could actually turn the car into energy. Einstein cannot be held responsible for this. Newton, who discovered law governing how an apple falls to the ground could not ultimately do anything about an airplane falling from the sky when the engines fail. On the other hand, Einstein, and not many people know this, was one of the key co-initiators of the American atomic bomb programme. In August 1939 he signed a letter to incumbent US President Roosevelt urging him to launch an American nuclear programme. Under no circumstances should the Third Reich get hold of the bomb before the Allies.

The explosive force of an antimatter bomb would be far greater than that of the Hiroshima and Nagasaki bombs taken together. Nevertheless, the danger of a nuclear holocaust will overshadow the likelihood of an apocalyptic battle of antimatter weapons for a long time to come. For antimatter is not biodegradable and even the artificial production of very small amounts is a very time-consuming and energy-intensive undertaking given the current state of the art. The storage and preservation of antimatter is still at a very early stage and the possibility of doing so is very limited. For this to happen antimatter must be separated from matter with strong electromagnetic fields in a kind of electromagnetic bottle to prevent an immediate reaction. No other method of storing antimatter is yet known. In principle, this method only works with positrons (antielectrons) and antiprotons because they have an electric charge and can therefore be influenced by an electromagnetic field. In addition, the method works only as long as the number of antiparticles is kept within modest limits. Once a critical number is reached, the interaction-related repulsive forces among the equally charged particles become so strong that the electromagnetic field is no longer sufficient to keep them in check.

Antiatoms such as antihydrogen or antineutrons[48] cannot yet be stored at all via electromagnetic fields since these antiparticles have no or only a neutral electrical charge and do not therefore react to electromagnetic fields.

The release of energy from antimatter and matter is in principle nothing more than an optimal nuclear fusion with maximum energy conversion - and definitely one of the cleanest ways of

---

[48] Antiparticles are generally distinguished because of the inverse electrical charge of their particles. The neutron and antineutron have the same charge. Nevertheless these are two different particles because the free antineutron (i.e. the antineutron not bound in an atom) disintegrates into a positron (antielectron) and an antiproton, but the free neutron disintegrates into a proton and an electron.

producing energy. Almost 2500 kilograms of antimatter (and 2500 kilograms of matter) would be enough to power the whole world for a year. However, as long as the artificial production of antimatter exceeds the energy potential of the reaction with matter, corresponding power plants or reactors are of course pointless. A natural source of antimatter that could be exploited with relatively little effort, like a conventional coal or oil deposit, would be the ultimate solution to the expanding energy problems of the world because one kilogram of antihydrogen combined with one kilogram of matter provides more energy than five hundred nuclear power plants in one year. Antimatter is certainly not found in any mountain or under any sea. There is, moreover, little hope of breaking down antimatter on a distant planet. For even the smallest contact with the stacked matter would result in a powerful reaction. But if one day it is possible to extract antimatter from a natural element or process, there would be nothing to stop its civilian and military use. Because of this, even a technology for storing antimatter would be superfluous since the antimatter could be produced directly in a reactor and converted to energy.

3.9 The Age of Antimatter

The most utopian plans in the drawers and archives of the American space agency NASA could be realised with the tremendous energy potential of antimatter. The conquest of the solar system. The colonisation of other planets. Intergalactic missions or the creation of extreme cosmic phenomena such as wormholes in the laboratory. The springboard to the future is hidden in antimatter, in the 22nd century, in the technology of the third millennium. Its energy potential holds the solution for breaking through a number of earthly blockages, blockages that have prevented the advance of technology into completely new spatial and temporal spheres.

Antimatter is an extremely compact source of energy. One gram of antimatter (about the weight of a strand of hair) contains as much energy as 23 fully fuelled space shuttles. Spacecraft with an antimatter drive would arrive on Mars or the most remote planets in our solar system in just a few weeks. It would even be conceivable for people to venture beyond the horizon of unmanned probes into the unknown depths of space for the first time. Strange stars and galaxies would be within our reach. The mastery of antimatter would overcome boundaries that would move humanity into a new era, a new epoch. Antimatter could even become the oil of the third millennium. The engine of the fourth industrial revolution. The revolution that lets us populate space.

The potential of antimatter is no longer a secret and certainly not a pipe dream that exists only in the minds of futurologists and do-gooders. The seriousness of the situation can be seen in the example of Kenneth Edwards, director of Englin Air Force Base in Florida. Edwards attracted attention in March 2004 when, in addition to numerous lectures, he published an interesting document[49] on the work of the US Air Force. It included concept sketches and plans for antimatter engines and novel military aircraft that would permit excursions to Mars. Edwards put the cost of a prototype antimatter-powered spacecraft propulsion system at $ 2 billion. Two billion US dollars. A relatively small amount considering that the stealth bomber, invisible to radar, cost an

---

49

www.niac.usra.edu/files/library/meetings/fellows/mar04/Edwards_Kenneth.pdf

estimated unit price of around $ 2.2 billion. Or the Apollo missions, which cost Americans $ 20 billion in the 1960's.

The first prototype should be available around the year 2020[50]. A bold plan. What touches the stars on paper comes right back to reality with a jolt if you want to fuel Edward's drive. The current situation is similar to a scenario in 2100, when all the oil reserves in the world will be more or less exhausted. Although every engineer knows how to screw a dirty diesel car out of the old blueprints, it is just the diesel that is missing. The world's most gifted researchers, with the most advanced equipment in the world, can produce fuel of the next generation only with great difficulty and in extremely small quantities. In addition, antimatter can hardly be stored for a long time, which inevitably leads to an uncontrollable reaction. As the oil pumped out of the earth in huge quantities with drilling rigs and derricks, the next era will not dawn until a method of extracting antimatter efficiently has been found. A method of producing the substance sufficiently and, where appropriate, a method of storing the substance. Like oil in a barrel.

But how would an antimatter engine work by 2020 if today's technology is not even enough to extract antimatter from the particle accelerator? Where would the fuel, the energy, the antimatter be obtained to operate such an engine at all?

Perhaps Edwards has misjudged the schedule, as we have already become used to among his contemporaries. The past provides only too many examples of technology and revolutionary projects that have turned out to be merely hot air. That is how the second millennium passed without NASA having undertaken a manned

---

[50] The document dates from 2004.

mission to Mars. Or nuclear fusion, which was to have replaced nuclear power plants, expected to be market-ready for the last fifty years or in fifty years time.

Maybe Edwards knows more than he lets on in his document. This is indicated in what is at first sight an inconspicuous note on the penultimate page, where he briefly states that the American Air Force is already researching revolutionary technologies to meet future energy needs. For this purpose experiments were under way which served to support the established theories.

What can be concluded from this?

A harmless indication that the US military still wants to tackle climate problems with clean energy sources after the launch of the Kyoto Protocol? Is the US Air Force looking for an environmentally friendly replacement to meet the shortage of fuel such as petrol and diesel?

Certainly not. Or at most enough to convince even Left-Green politicians of the need for military spending. This suggests that the US Air Force is taking anti-matter technology seriously and is not just limiting its efforts to conceptual outlines. Without denying it, a process of releasing antimatter from an earthly accessible material could be the foundation for industry and technology over the next decades and centuries. To achieve this, a physical method would have to be discovered that enables antimatter to be generated from elements or processes available on Earth. The periodic table of elements currently consists of 118 elements which are not necessarily natural but could be detected in the particle accelerator. All known elements with an atomic number[51] higher than 94 can

---

[51] The atomic number indicates how many protons are present in the

only be formed in the particle accelerator. Their detection was not possible in Nature due to the very short decay time. So far, we do not know much about the elements beyond the 118th element. Theoretical considerations show that there must be many more extremely heavy elements. With increasing atomic number, however, the elements would be increasingly unstable and thus harder to detect and generate. It is conceivable that elements with an atomic number greater than 118 occur in nuclear fusions, where very high energies and temperatures of a few million degrees occur or exist in supernovas or neutron stars. However, these elements cannot yet be generalised to determine whether they are radioactive and thus unstable, or whether there are certain islands of stability, that is, certain atomic number anomalies in which even very heavy elements prove to be stable. Such elements with a decay time of more than one second would be the prerequisite to using them in practical applications.

We cannot say anything about anomalies at very high atomic numbers. We do not know whether, for elements with atomic numbers over 500, relativistic or quantum mechanical effects cause at best a hitherto unthinkable stability or other peculiarities. In elements beyond the presently known periodic table it may be possible to find elements that are relatively stable and release antimatter on disintegration or in combination with other elements. If we could prove such a process it would be possible, at least on paper, to make antimatter energy efficient. Storage could be omitted if the antimatter is generated and used directly in an engine or generator. Little matter from these superheavy elements could

---

atomic nucleus of the element.

generate enough antimatter to move spaceships with a comparatively light tank to distant planets. This opens up new possibilities for space research that are inconceivable using established solid-fuel rockets.

However, we do not yet know which properties develop elements with very high atomic numbers and whether it is even possible to refine antimatter in some way from materials available to us. That does not mean that it is impossible.

3.10 The absolute zero point

Heat leads to drought and dryness, allows the desert to conquer populated areas and dry crops. Cold freezes life in the ice and preserves the past as a geological museum. Warmth and coldness accompany humanity like the moon, which travels orbits the earth like a faithful companion.

The science of heat and cold, thermodynamics, attracted scholars as early as the end of the Middle Ages.

As the stakes burned, the academic elite were puzzling over the nature of temperature. The French physicist Guillaume Amontons, who was born in the City of Love on August 31, 1663, was also attracted to thermodynamics. In 1699, when he linked the volume of a gas to its temperature, he started the ball rolling to the extent that this subject is still fascinating to this day. Amontons had discovered that a gas takes up more and more space the hotter it gets. The temperature had to be related in some way to the volume of a gas.

In fact, the first railway workers quickly realised that all materials expand as soon as they are heated. This is how rails and bridges twist into profile when the blazing sun burns mercilessly. Or take the LHC particle accelerator at CERN, which contracts a few feet when brought to operating temperature (about minus 271 °C). Heat and cold can cause serious damage if the designers have not given sufficient consideration to these phenomena. This is also the reason why modern concrete bridges always leave a margin at both ends in which the bridge can expand. In the summer, for example.

However, Amonton's discovery did not prompt the rulers of that time to hunt down an emergency team of architects, technicians, and premature engineers across Europe to redevelop all bridges and structures. Rails were known mainly from gold mines but some years would elapse before James Watt introduced his steam engine. The contemporary bridges were also made of wood and were therefore less rigid and subject to thermal differences, at least as far as expansion is concerned.

The matter became much more enigmatic when Amontons and his contemporaries wondered how this thermodynamic phenomenon was to be understood in reverse. What happens to a material that keeps on cooling? What will become of your frozen foods if you let your freezer cool down?

There were two possible explanations. There is an absolute zero of temperature, a "cold" that you can never go below. A temperature point at which the volume of the gas is zero, your Hawaiian pizza shrinks in the refrigerator. Or alternatively there is explanatory approach number two where the relationship between volume and temperature applies only to gases and not to liquids or even solids.

Not until the French Revolution "beheaded" the monarchical social order, James Watt presented his steam engine and mobilised the textile industry to mechanical manufacturing did William Thomson find the answer. In the year 1848, when the gold rush was about to explode in California, Thomson stated that the energy lost from the material was critical to this issue, not volume reduction. The colder a material, the less energy it has. You can cook with hot water, prepare spaghetti or burn your fingers. Hot water is energetic whereas cold water has lost energy, comparatively speaking. At absolute zero, the volume of a material is not infinitely small, but rather the material is "without energy". Thomson therefore proposed a new temperature scale, the Kelvin.

The Kelvin has no negative values, since it begins at zero Kelvin or minus 273.15 degrees Celsius. This is the coldest possible temperature value in our universe. In no experiment in the world could this ice barrier ever be crossed. No substance and no matter can become colder than this chilly constant of Nature. Absolute zero is a barrier that will not let anyone pass. A railroad crossing that remains closed to us forever. A limit of accessibility, just as the speed of light. No massed particle can fly faster than light, no matter what energy it consumes. Similarly, no particle can fall below the absolute zero. Zero Kelvin is the absolute limit of cold.

The cause of this leads us into the veiled world of quantum physics. Heat and cold are nothing else than statements about the kinetic energy of atoms. The glowing hotplate or the surface of the sun is usually considered hot. The Arctic ice or the climate on the summit of Mount Everest, however, is on the other hand considered cold. What Nature understands by heat and cold is merely kinetic energy, however. The warmer an atom is, the faster its particles move and vice versa.

This fact can be illustrated with a sprinter in winter sitting on the edge of a race track waiting in the frozen grass for the start of the race. He is freezing and his body contracts. As soon as the starting signal comes, he starts running. His body is stressed and requires a lot of energy to maintain a fast pace. As long as he runs he is warm, he may even start to sweat. But as soon as he stops again and therefore runs out of kinetic energy, his body cools down again and he starts shivering.

When a substance is cooled to absolute zero, it loses all its energy. The particles are stationary in this extreme case. This is the assumption one would intuitively make. The reality, however, turns out to be a little more complicated because absolute zero is one of the strange barriers of the universe that cannot be overcome or reached. Beyond this barrier a world with the natural laws known to us would be very equivocal and difficult to describe. The absolute zero point is another borderline phenomenon on the horizon of explicability.

Quantum physics and relativity play a pioneering role when it comes to explaining such phenomena. But in order to be able to fully understand these marginal phenomena, we would probably have to resort to a hitherto unknown theory.

In fact, the particles of a material can never be completely shut down. Nature cannot be seen in the cards. Quantum physics prohibits anything from remaining at absolute rest. Even at absolute zero, the particles of a substance cooled to zero Kelvin still move with a residual energy. A residual energy that cannot be measured by any thermometer in the world, but must exist according to the laws of quantum mechanics. This phenomenon is often referred to as "zero point energy" and is due to the

Heisenberg uncertainty principle. An important principle of quantum physics, hence the place and the momentum[52] of a particle are never measurable simultaneously. Not because the equipment or measuring technology would not be able to do so, but because nature is "built" the way it is. But if the particles of a material are in absolute calm, the location and the momentum would be measurable at the same time, which would violate the Heisenberg uncertainty principle and hence an important principle of quantum mechanics.

But it is quite possible to cool a matter to almost zero Kelvin. Just as it is at least theoretically conceivable to accelerate a spaceship arbitrarily close to the speed of light. Researchers from all over the world have already managed to reach the low point of cold to a billionth of a degree[53] and observe all sorts of peculiar phenomena. Phenomena which occur again and again if you try to catch a glimpse behind the barriers of nature.

Let us recall: The closer a space ship approaches the speed light the stranger the effects will be. Effects like time dilation or length contraction. Effects that are completely foreign to us in everyday life. Also, the zero seems to be in an area of anomaly, anomaly denoting the strange properties that materials in this temperature range may assume. Take superconductivity, for example.

Superconductivity is a physical effect which in the future could allow humanity to make tremendous progress in the energy and

---

[52] Momentum is defined as the product of mass and speed.

[53] However, the zero point is still unreachable even at a great distance, for every closer approach to the zero point requires an increasing amount of effort, just as each acceleration towards the speed of light requires an increasing amount of energy.

technology sectors. In extreme cold, superconductors can transport and store electricity without loss of energy because the electrical resistance (which leads to heat, for example) disappears at a critical temperature. This is the dream of every energy company, environmentalist and supporter of portable multimedia devices. Until now it has been impossible to conduct power without loss via high-voltage lines or even store it. Although accumulators and batteries supply Notebooks and MP3 players with electrical power this arises from chemical reactions and is not stored as "electric current". Research into superconductivity could someday make the dream of a true "electric battery" or lossless power transfer a reality.

The problem is that substances become superconducting only at very low temperatures. Lead or tungsten (from which the wires of conventional light bulbs, for example) develop superconducting properties very abruptly close to absolute zero. The transition temperature[54] of lead is around 7 Kelvins, that of tungsten well below one Kelvin. This is also the reason why superconductors in everyday life so far have led a rather quiet existence. Cooling a 100-kilometre power line to over minus 270 degrees Celsius is no easy task. The possibility of using superconductors for this purpose is arousing considerable interest in research. They are ideally suited for generating strong magnetic fields and are used, for example, in the LHC particle accelerator at CERN in Geneva. There have been recent discoveries of high-temperature superconductors, i.e. superconductors that lose their internal electrical resistance even with relatively little cooling. Their temperature was still around

---

[54] The term "transition temperature" is used because the substances suddenly lose their electrical resistance from a certain temperature and therefore become superconductive.

minus 155 degrees Celsius. However, there is every hope that one day superconductors will be discovered which will exert an effect even at room temperature. That would be a huge technical advance. From now on electrical power could be transported and stored with virtually no loss. But no one knows whether there are materials that become superconducting in these high temperature ranges. The theory is still rather inconclusive when it comes to explaining this phenomenon, which is why no maximum limit can yet be determined for superconductor properties.

In recent years spectacular reports have repeatedly emerged where superconductors were said to have an anti-gravitational effect. With superconductors, it should therefore be possible to shut off gravity and hence overcome the power of attraction and space-time curvatures. If these reports, which are not without controversy, were to be confirmed and the experiments were actually to be reproduced, this would be an absolute sensation. So far, there is no known mechanism that could isolate gravity. Gravity is the only one of the four known basic forces of Nature that cannot be isolated, either with an arbitrarily thick steel plate or with an entire star (the gravitational field would be additionally reinforced by its mass). Gravity cannot be tamed in four-dimensional space-time. Except with fast rotating superconductors, as some researchers claim. The evidence of this, however, is as elusive as a conclusive physical explanation.

On the other hand, the "Bose-Einstein condensate" predicted as early as the 1920's has been clearly confirmed in experiments. At temperatures near absolute zero some particles fall into an extreme state of aggregation in which they can no longer be distinguished from each other. The particles lose their characteristic properties and appear as one united "super atom". There is a degeneration of matter, rather like when an individual joins in the jubilation of the fans when the when the football team scores a goal at their home stadium. However, the "Bose-Einstein condensate" can only be observed in "bosons", a very specific form and type of particle.

Incidentally, Nature does not just have a lower temperature barrier. Purely in theory, there is also a maximum heat or rather heat that can reach matter. This results from the Planck temperature and is an exorbitant $1.4 * 10^{32}$ Kelvins (spelt out: the number 14 with 31 zeros). Without having to know how to calculate this value it may be concluded from simple considerations that there must be an upper temperature limit. As is well known, temperature is nothing but kinetic energy. The particles cannot move faster than light. Consequently, in principle, no temperature higher than the value at which the particles move at the speed of light is possible (particles that have mass can never reach exactly the speed of light). This is only a minor mental game and without considering other relativistic or quantum mechanical effects.

Incidentally, even in the most remote areas of the universe the temperature is approximately three Kelvins, which is relatively well above absolute zero. This fundamental temperature of the universe is probably due to the background radiation that emerged at the birth of the universe. It is still unclear how the temperature could be distributed evenly - even in areas of space that are too far away to have had enough time to exchange information with other areas. The age of the universe is estimated at about 13.7 billion years. If a galaxy A is seven billion light-years from Earth and one galaxy B is also seven billion light-years in the opposite direction, the total time since the big bang is not enough to allow communication or information exchange between the two galaxies. All information can propagate at a maximum of the speed of light, which is why the light takes seven billion years to reach the eye of the beholder on Earth from galaxy A seven billion light-years away. Nevertheless, in the entire observable universe practically the same physical properties appear to dominate, which can hardly be explained by coincidence alone. This inconsistency in the cosmic standard model is called the horizon problem and provides an indication of the incompleteness of the general theory of relativity. One approach offers the theory of cosmic inflation whereby the universe expanded shortly after the Big Bang at many times supraluminal velocity[55] and light was entrained almost in the wake of this expansion.

Background radiation has therefore become evenly distributed throughout the universe, which explains the small local differences that can be observed today. Another theory was postulated by the Portuguese Magueijo and the American Andreas Albrecht in 1999, in which they explain the horizon problem with a variable speed of light. Here the speed of light in a vacuum is variable over cosmic time scales, and in the early phase of the universe was about 60 times that of today's measurement. This allowed even the most remote regions in the Universe to communicate with each other, distributing the background radiation evenly. However, experts are rather critical of this theory as it questions the constancy of the speed of light as the basic postulate of the theory of relativity.

Whether or not the natural constants have changed in the history

---

[55] Einstein's theory of relativity does not provide information on expansion at supraluminal velocity, but it is conceivable that space-time expanded as a structure of the universe at supraluminal velocity (viewed from our perspective) without violating Einstein's postulate.

of the universe can perhaps be answered by string theory, a theory of everything candidate, namely when we succeed in determining the constants of nature not only by measurements, but by attributing them to a superordinate principle from which they can be derived. We would then have understood the constants of nature and we could probably explain why, and how often, for example, the speed of light has changed in the past[56].

3.11 The Higgs particle

In the 1960's and 1970's, science succeeded in formulating a theory that takes into account three of the four fundamental forces of nature and the special theory of relativity. This theory is called the Standard Model and includes just about all the essentials we need to know about the microcosmos in physics.

The standard model is not a universal formula. For this to be so at least the general theory of relativity, and hence gravity would somehow have to be formalised. But outside a universe of ten, eleven or even twenty-six dimensions, this seems to be an impossibility. Although the standard model describes the processes in the microcosmos very precisely, it is not enough to describe and understand all aspects of physics. In particular, the standard model is based on eighteen parameters that had to be determined experimentally and injected into theory. These values cannot be deduced or explained on the basis of the model. Similarly, as in the theory of relativity, certain natural constants such as the speed of

---

[56] In this case the constants of nature, such as the speed of light, would only be constant within a short timeframe, but not at cosmic distances, and in this sense they are not universal constants. However, nobody yet knows whether the speed of light has changed at all.

light have so far only been experimentally determined and cannot be calculated from the theory. This makes the standard model flexible and elastic and can be adapted relatively well to the experiments. This is a scientifically unsatisfactory condition in the long run and could give the theory the reputation of having been thrown together from experimental data and mathematical sophistry. But for now we have to accept that we cannot derive this natural constant. However, it is conceivable that these natural constants necessarily result from the formalism of a world-formula theory.

The standard model quantizes power transmission. That is, the interaction of the electromagnetism, the weak and the strong force are transmitted only in whole (discrete) packages. The standard model also has three different types of particles: force particles, matter particles, and the ominous Higgs particle.

Force particles transfer the three non-gravitational basic forces. The photon, for example, is the force particle of electromagnetism and transmits the attraction that two unequal charged poles exert on each other. It is believed that gravity is also transmitted by a force particle, the graviton. The graviton as not yet been detected. Insoluble contradictions have hitherto stubbornly prevented gravity from being integrated into the standard model. In the microcosmos the influence of gravity is extremely weak and is therefore negligible, just as the effects of relativity or quantum mechanics are barely noticeable in everyday life.

Particles of matter represent the elementary particles familiar to us, such as protons, neutrons, electrons or even the muon. For every particle of matter and force there is an antiparticle consisting of antimatter. So far researchers have been able to detect all the predicted particles of the standard model - except for the Higgs particle.

The Higgs particle stems from the bold attempt to explain where particles have their mass and why the particles have such different masses. An electron is about two thousand times heavier than a proton, although both have exactly the same electrical charge. The photon, in turn, is massless and the we are still not sure whether the neutrino may have an infinitesimally small mass. What sounds rather clumsy could perhaps be the holy grail of the Standard Model because the standard model requires massless, punctiform particles. With elementary particles that have mass its formalism would collapse.

In order to overcome this dilemma different scientists independently came up with the idea, in 1964, of introducing a mechanism that gave mass to even elementary particles through a force field. The British physicist Peter Higgs published his theory first, which is why the particle was named after him. He was still unaware of the significance of his publication, so he self-critically wrote to a student: "I have discovered something completely useless".

However, the so-called Higgs mechanism is by no means that useless. In the Higgs mechanism, he described how the particles gains their mass. For this purpose he introduced a background field, the Higgs field, which spans the entire space-time and fills the vacuum like a viscous fluid. The particles are slowed down by the interaction with this Higgs field, causing mass or inertia. It's like slowing down a speedboat as it moves thorough a swamp. In Newtonian mechanics inertia, the resistance of a particle against a change of motion, had already been equated with mass. Einstein extended this basic assumption to the principle of equivalence, in which postulated that acceleration and gravity are similar in nature. In doing so, he presupposes the validity of the principle of equivalence, but without being able to explain the equality of inertia and mass. Do two such fundamental properties of physics as inertia and mass arise from the Higgs field? Should it be possible for the first time to decode such a fundamental particle property?

A government party, which, exceptionally, has no intention of wasting taxpayers' money, illustrates how the Higgs mechanism works. In a large ballroom numerous representatives of a ruling party are bustling around. When the president of your own party enters the hall, a group of people form around her, all wanting to greet her or speak to her. Consequently the president makes slow progress. If a representative of the opposition enters the room, everyone present withdraws since nobody cares about the political opponent. The member of the opposition can therefore move freely and unchecked throughout the hall. The president in this example corresponds to a proton. The proton is strongly braked by the Higgs field and thus receives a large mass. The member of the opposition is the photon. The Higgs field exerts no braking effect on the photon, which is why it remains massless. The electron could be thought of as a waiter entering the room with a plate of fine biscuits. Of course he attracts less attention than the president, but is still besieged by some hungry politicians. He is thereby slowed down in his movement and receives his mass in this way. An electron is therefore much lighter than a proton because it is less sought after or is slowed down less in the Higgs field. This braking effect is what gives particles their mass.

With the Higgs mechanism, the dilemma of massless particles has to be solved. The ball now lies in the court of the experimental physicists, who first have to find out if this mechanism exists in Nature at all, for example, an experiment in which the Higgs particle is detected in a particle accelerator. CERN is currently running the LHC experiment, with the detectors "Atlas" and "CMS" providing clues to the existence of the particle. In fact, according to initial findings, there is an unusual frequency of particles within a certain energy range in which the Higgs particle might fit. If it turns out that the Higgs particle does not exist, that would be a big blow to the standard model. The question of the origin of the mass would have to be re-examined. But even if the Higgs particle does exist the standard model would still have to answer a number of outstanding questions. So it is still unclear why the fundamental fundamental forces are so different. Or how the standard model can be combined with theory. Or whether and how the eighteen free parameters can be predicted from a general theory. Or how the different masses of particles can be explained.

The Higgs mechanism explains how particles receive their masses, but not why a proton is more massive than an electron. All we know is that the proton in the government hall is more coveted than the electron, but not why. Likewise, the importance of the Higgs mechanism is not yet clear. The Higgs particle is, so to speak, the only interacting particle that does not belong to any natural force. How is this special role to be understood? Is the Higgs mechanism something like nature's fifth fundamental force? Or do space-time and Higgs field fuse together, so that the mass becomes a consequence of space-time? Would it be conceivable that the Higgs Field is nothing other than a property of space-time that has not been understood or recognised as such? In this case, would it be conceivable that gravity cannot be reconciled with the standard model because it is already contained in some way in the standard model through the Higgs field? Would gravity therefore be nothing but a curvature of the Higgs field and the rest mass to a certain extent the basic acceleration or the basic resistance, which is opposed to an unaccelerated particle by the Higgs field or space-time?

From the equivalence principle we know that gravity and acceleration are equivalent in their effect and in the reference system, and cannot be distinguished on the basis of the reference system. Would it be conceivable to visualise gravity and acceleration as a bodyguard which pushes the president through the room? As a result she is accelerated but the politicians surrounding her do not yield voluntarily and therefore counteract the acceleration. This illustrates approximately the effect which we in physics describe as inertia. Instead of the bodyguard, a tilting of the banqueting room may be envisaged in order to visualise gravity. Here the president and the politicians surrounding her slide "downwards" along gravity. The inertia or mass remains the same as the politicians slide with the president.

Perhaps the key to unifying the standard model with the general theory of relativity lies in an understanding of space-time. Perhaps the Higgs field is really nothing more than a hitherto misunderstood consequence of the structure of space and time. Thus the missing piece of the puzzle might be hidden in this structure and a quantisation of this structure, space-time, might perhaps be the correct approach for advancing the theory. But let's leave this speculation to the theory of everything candidates. String theory and quantum-loop gravity make every effort to unite the four basic forces. We might also be anxious to evaluate the data from the LHC experiment. Then maybe we'll learn if the Higgs mechanism actually exists in nature. This will deciding factor in determining whether we are or are not in the Higgs mechanism.

4. The theory of everything

The search for the blueprint

The physicists have a common dream: to find a formula from which all natural laws can be derived. A formula that explains the lever principle as well as the eerie long-range effect. Albert Einstein dedicated a considerable part of his life to the search for such a blueprint on which all laws and forces could be based, but even he failed in this bold project.

The problem is that quantum and relativity physics by no means compatible in some situations. The quantum theories may very well describe the behaviour of particles and forces on a small scale. The theory of relativity in turn explains gravity and the nature of space and time. But there are some stubborn cosmic phenomena that involve both gravitational and quantum effects. Inside a black hole, for example. If we try to describe this phenomenon based on today's physics we get fractions with a zero in the denominator, a result which is not permitted in any pragmatic mathematics. In other words, a result that is infinite, which in most cases indicates the incompleteness of the underlying theory.

The fundamental question, of course, is whether gravity can even be combined with quantum physics in a single theory. In fact, the two theories are quite alien. Thus, the theory of relativity explains gravity with a gently curved, consistent space-time, whilst quantum mechanics wild and chaotic activity prevails. The closer one looks, the more blurred and confused the quantum foam becomes. How can the calm, elegant theory of relativity be brought into harmony with the wild quantum nightmare? And why are the physicists ever so eager to squeeze the theories into one overarching explanation?

First, let us address the final question, the basic motivation that persuades physicists from all over the world to devote their lives to the development of a theory without being sure of ever finding it. The physicists have a theory in the standard model that enables them to calculate most of the events in the quantum world fairly accurately. Over the decades they have even succeeded in combining three of the four basic forces of nature (the weak, strong and electromagnetic force) with the special theory of relativity. Presumably, these forces combine at very high energies into a single superpower and therefore objectively share a common scientific foundation. Despite all efforts it has not yet been possible to integrate gravity into the standard model, or in other words: to quantify gravity, i.e. to find a smallest "set" of gravity. The scientists suspect the graviton as a hypothetical, massless particle that transmits gravity, but so far they have been unable to prove it. The graviton would therefore be the transmission particle of gravity, as it is the photon of light or electromagnetic waves. But it is not only the quantisation of gravity that is an unsolved problem. In the current standard model there are eighteen parameters whose values have been determined in experiments and are not evident from theory. To date we have not found any explanation for these eighteen natural constants. Although we know their experimental values, such as the speed of light or the gravitational constant, we cannot explain why they assume exactly that value and no other. It follows that there must be a higher theory which reduces at least some of the natural constants to a deeper principle and thus answers the question of why. Discovering this theory, and thus solving some of the most fundamental question marks in modern physics, provide the incentive and motivation of many researchers to engage with world-formula theories.

The history of science teaches us that a new chapter of physics is often written when people believe they have unlocked the ultimate secrets of Nature. In fact, there are really only a few questions that are not answered in the current standard model. But these questions are so fundamental that their answers can only be surmised in a theory that is beyond the scope of the standard model. A theory that includes modern physics as a limiting case. Just as Newtonian mechanics has come to fruition as a limiting case for slow velocities in modern physics, modern physics may someday become part of the universal formula.

But what is this theory of everything? What is the theory that stands above the pillars of modern physics? And what does the theory of everything tell us about the nature of the world, matter, energy and forces?

The so-called theory of everything is often referred to in professional circles as the attempt at a quantum gravity or unifying theory. Over the past two decades, two promising theory of everything candidates have emerged from the research strategies adopted. On the one hand the string theory, which postulates vibrating threads as basic building blocks of the elementary particles and a universe with at least ten dimensions. On the other hand, loop quantum gravity, which considers space, time, and the entire universe as a vast network of small quantum. We focus on string theory below.

But before we plumb the depths of string theory, let us first and foremost answer the question of what the theory of everything is. The theory of everything is a currently hypothetical theory that explains and links all known physical phenomena. Hypothetical because it has not yet been proven (or refuted). This involves, among other things, bringing gravity and quantum theories together under one roof. The theory of everything does not necessarily have to be a single formula, but a single set of formulas describing all physical phenomena. This would then also enable us to understand and calculate the inner life of black holes.

However, the theory of everything is neither the last word nor a hotchpotch of universal knowledge. The theory of everything is said to shed light on on what are today inexplicable phenomena. For example, it answers the question of what dark matter is, why the elementary particles have their particular mass and why gravity is much weaker than all other basic forces of Nature. But it is wrong to think that the theory of everything would somehow open up the plan underlying our existence. Although the theory of everything explains all physical phenomena in a single theory, it does not say anything about processes of psychology, thinking or feeling. The complexity of these processes alone is such that we cannot possibly calculate them, even if we have a knowledge all the necessary formulae. So today we can make statements about the behaviour of the planets or the particles, but these predictions are often only approximations or limited to an isolated system. As soon as even a small number of systems interact, for example four or five celestial bodies, the calculations are already complicated. But in complex systems like the human body, millions and billions of particles interact, making it impossible to calculate the organism precisely. In addition, the universal formula must take into account the uncertainty principle of quantum mechanics within the meaning of the correspondence principle. It follows that in principle a system can never be calculated exactly. Measuring one property with a certain accuracy always causes another property to blur by that same precision.

So let us sum up: The theory of everything is a superordinate theory that summarises all the basic forces of Nature and explains all known physical phenomena from a single formula. The theory of everything does not explain why we can think, feel or be creative. Nor can the theory of everything predict the future or calculate complex systems such as the human organism. The reason for this lies on the one hand in the randomness of microscopic systems, on the other hand in the complexity of the resulting calculations.

Einstein was not the only scientist who failed in the theory of everything problem. Renowned physicists such as Werner Heisenberg, Theodor Kaluza and Oskar Klein have also embarked upon the search for the coveted formula. But none of them has found anything. At least not completely. Kaluza and Klein made significant progress in the 1930's. They found an original approach for combining electrodynamics with gravity. To do so, they extended space-time by a fifth dimension. At the same time, quantum mechanics developed very rapidly and in turn transferred electrodynamics to quantum electrodynamics, that is, into a quantum theory. The Kaluza-Klein theory, however, could not be quantised, which is why it soon fell into oblivion, even though Einstein personally praised its elegance and boldness. It was almost half a century before the Kaluza-Klein theory was rediscovered. Or at least the ingenious idea of assigning basic forces of nature to additional dimensions. Even the theory of relativity had pursued this approach, in which it attributed the gravity to a curvature of the four-dimensional space-time. What was revolutionary about the Kaluza-Klein theory was the idea that there could be dimensions outside our perception. We know the dimensions of space-time from our everyday life. On the one hand we have three space dimensions (up / down, right / left, front / back) and on the other hand we have a time dimension. Kaluza and Klein now suspect that the fifth dimension is wound up like an extremely small bundle of twine. So small that it remains invisible to us. This circumstance can be illustrated with the following example: If you hold a garden hose in your hand, you perceive it in three dimensions. To you the hose has a width, a length and a height. From a distance, however, the hose only appears to you as a one-dimensional line. You observe the same optical illusion at night when you look at the stars in the sky. The stars appear to you as many small, luminous points. If you were to fly to one of these stars, you would realise that they are huge celestial bodies. A wrapped dimension is invisible to us because it is compacted in the order of Planck length. The Planck length is the smallest unit of length for which our laws of nature are still valid. Everything below that unit of length would instantly collapse into a black hole. We do not have anywhere near the technical prerequisites to observe such a dimension directly. Not even with the most up to date particle accelerators. But it is still a fascinating thought that our world inside is ticking very differently than we experience in our everyday lives. Not only scientific, but also philosophical questions are connected with this. If you look at a grandfather clock and follow the hands, you will no doubt agree that the small hand indicates the minutes and the big hand the hours. So that is what we know as time. You know from your own experience that a watch shows the time. But what if it is just a huge coincidence that the clock shows the time but in truth serves a completely different purpose? If the world consists of numerous dimensions that we do not know about, can our everyday experiences tell us anything about the nature of the world? In everyday life we experience the world as consistent. Time passes continuously and inexorably. If you drop your briefcase, it falls to the ground. Even if you drop it a thousand times, it will still fall to the ground. But in the microcosmos we experience a completely different "restless" world. Now that quantum mechanics has celebrated its debut, we know that probabilities, coincidences, and unpredictable events shape this part of the world. On the one hand, we have our everyday world that we know, and on the other, a wild and random microcosmic world from which the cosmos and the universe are built. So what is the reality? Or to put it another way: what if the wild world of quantum was our everyday life? The string theory and all theories claiming to be a unifying formula ("theory of everything ") deal precisely with these questions. In principle, string theory explains why quantum physics and the theory of relativity differ so much. And how is it possible for us to know the world from our everyday experience as consistent, even though the building blocks of the same world are anything but consistent. The Kaluza-Klein theory could not solve this problem because it was not compatible with quantum theories.

4.2 The String Theory

In one of the first chapters of this book we learned that matter consists of many small particles. Its kitchen table is not just top and feet, so to speak, but a collection of billions and billions of tiny particles. In ancient Greece, the philosopher Democritus suspected early on that all matter consists of atoms - of the smallest, inseparable elements. Only at the beginning of the twentieth century did physics discover that atoms in turn consist of electrons, protons and neutrons, and hence are not quite so elementary as initially assumed. Added to this was the discovery of particles of cosmic radiation (muons) and neutrinos. Five elementary particles were therefore known. Over the years, a heated debate arose over the question of which particles can actually be described as elementary particles because more and more particles were discovered. This gave rise to the assumption that there would be still smaller particles that would be really fundamental. For physics, it was considered unsatisfactory that the building block of matter should be a steadily growing zoo of particles. In 1964 the physicist Gell-Mann took the lead in predicting quarks, which in turn consisted of even smaller particles, of which, for example, protons and neutrons are composed. Five years later he received the Nobel Prize. Today we know of exactly three families of different quarks and a large number of different elementary particles that consist of them. In fact, in addition to the protons, neutrons and electrons, there are numerous other elementary particles that are composed of different combinations of the different quarks. The kitchen table consists of atoms, the atoms of elementary particles, the elementary particles of quarks, the smallest known building blocks of matter. However, quarks do not exist outside of particle bonds, which is why we may continue to designate neutrons or protons as elementary particles. But what on earth is the quark made of? And why are there exactly three quark families, and not 4, 5, 6, 7 or 8? And anyway: What are the real building blocks of matter, the very smallest particles that exist?

By the mid-1970's quantum mechanics was fairly well formulated and numerous results confirmed by experiment. As quantum mechanics seemed to be approaching full decryption, some physicists were looking for new areas of research. And they were to find something. The mysterious conflict between quantum physics and the theory of relativity remained unresolved. Both theories were experimentally very well confirmed, but they were in conflict with each other. Space-time, according to the Theory of Relativity, is flat and calm when there is no mass or energy present. It is like a sheet of paper lying on your desk. Quantum physics shows the exact opposite. As you continue to enlarge an area (which means smaller and smaller distances) it becomes increasingly blurred and the quanta move ever wilder (fluctuate). This so-called quantum foam is what has so far made it impossible to simplify the General Theory of Relativity and quantum physics[57].

Now was the time to tackle this enigmatic conflict. In search of creative approaches to solving this problem, some physicists have rebuilt the Kaluza-Klein theory. This had fallen into oblivion because it was not combined with the booming quantum mechanics. By now sixty years had passed and two hitherto unknown natural forces (the strong and weak force) had been discovered. Some experts now suspect that the Kaluza-Klein theory may have been formulated with too timidly. At that time, nothing was known about the weak and strong forces. Kaluza used a five-dimensional space-time to combine gravity with electromagnetism. He pursued the idea that electromagnetism is also characterised by a curvature of its own dimension, as gravity is characterised by the curvature of the four-dimensional space-time. The physicists now speculate that not yet aware of the weak and strong forces, he might have been too cautious in his dimensioning. But the combination of all the forces might possibly succeed if the theory were extended by some extra dimensions. A basic concept of string theory was born. Indeed, it was to be demonstrated that string

---

[57]   Meanwhile the special theory of relativity can easily be integrated into quantum physics. The problems only arise with the General Theory of Relativity which embodies gravity.

theory needs at least ten dimensions to deliver meaningful results. A result is useful if it provides a finite value. In quantum physics, we can predict the behaviour of particles or the occurrence of an event only with a certain probability. For a result to make sense in probability theory it must only assume values between zero and one, or between zero percent (never occurs) and one hundred percent (certainly occurs). In all previous attempts at unification there were always negative or infinite probabilities. These were clear indications that the theory was incomplete or incorrect. Only if one starts from the calculations of at least ten dimensions does the string theory yield meaningful results. For string theory to make sense, the universe must consist of at least ten dimensions. Namely from nine room dimensions and a time dimension. Without these extra dimensions, string theory gives us about as much as a car without a steering wheel. In the early 1970's an attempt was made to find a strong quantum theory. This force is one of the four basic forces, and put in simple terms, it holds together the atomic nucleus (protons and neutrons). An attempt was now made to describe the strong force as one-dimensional particles, so-called "strings". Until then, it had been assumed that the quanta are zero-dimensional point particles, thus electrons have no spatial extent, for example.

In 1974, this one-dimensional string approach only worked in a 10 or 26 dimension universe. Through the work and research of well-known scientists such as Joel Scherk, or later also Michael Green and John Schwarz, it became apparent that this "string" approach was not only a hypothesis for the formulation of the strong power but also an approach for unifying the quantum and relativity theories. In other words, a pillar of the theory of everything.

With the first so-called superstring revolution in 1984 Michael

Green and John Schwarz were able to prove that string theory leads to a ten-dimensional supergravity where there are only left-handed neutrinos (parity violation). Without wishing to go any further here, this discovery was a powerful boost to string theory research. Similarly, string theory has the "remarkable ability to predict gravity," as Edward Witten once said.

Although neither Kaluza-Klein nor the researchers of the strong force explicitly sought the universal formula, nor did they discover two of the fundamental elements of the resulting string theory. On the one hand the hypothesis of a ten-dimensional universe with wound up extra dimensions wound, and on the other hand, the strings as one-dimensional building blocks of all matter.

Now that we know how string theory came into being, let us look at the question of what string theory is exactly.

4.3 Strings - The smallest building blocks of matter

In string theory, the smallest building blocks of matter are no longer the quarks or elementary particles, but one-dimensional threads, so-called strings. These strings have a finite length, in contrast to the previous standard model, in which the elementary particles were considered to be zero-dimensional points (whose size was correspondingly infinitesimal).

The strings may be imagined as the strings of a guitar. When you play the guitar and pluck the strings you get different sounds. The different elementary particles and force transmission particles arise because of the excitation of the strings or the fact that the strings vibrate. The faster a string vibrates the more massive the elementary particle that it creates. All matter and all forces are ultimately due to this "simple" basic element, the strings.

As we discussed in the previous chapter, string theory requires at least ten dimensions (nine dimensions of space and one time dimension). The four dimensions of space-time are known to us, leaving six dimensions. Since we have not seen these dimensions before and they are not accessible in our experiments, we can assume that they are wound-up dimensions.

These six extra dimensions have now wound up like a ball of wool at any point in space-time. If you turn your head you move through millions and millions of these six-dimensional balls.

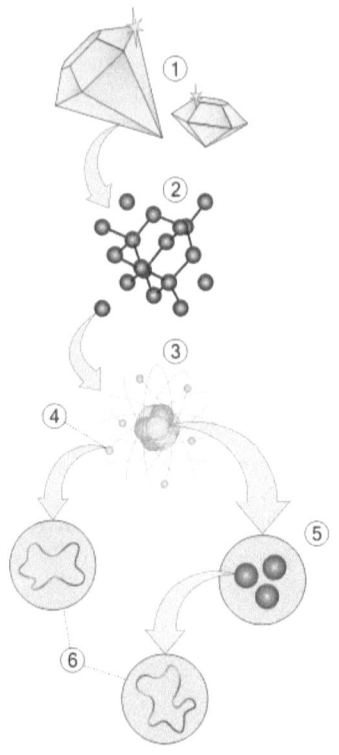

Legend

1. Matter
2. Molecular structure
3. Atoms
4. Electrons
5. Quarks
6. Strings

Figure 10   The fundamental building blocks of matter

But since these are very small and also everywhere in the entire room, you do not notice them. Every oscillation of a string always vibrates through all the extra dimensions. The way these dimensions are wound up and intertwined crucially determines the physical properties of the elementary particles that result from them, such as the particle mass or particle charge. If string theory succeeded in predicting the particle mass or some other fundamental property of an elementary particle, this would be a first significant indication that it is true. The question of why the elementary particles have a particular the mass and not some other is one of the greatest puzzles of physics. The standard model has no answer to this. Although it describes the interaction of the fundamental forces (except gravity) and the elementary particles, it refers to eighteen parameters that must be determined by experiments. From string theory, these parameters, as well as the masses of particles and many other physical properties that we previously took for granted (e.g., the speed of light, the value of the gravitational constant, or the charge of the electron), would have to be deducible as the logical consequence of a more fundamental principle. The prediction of a previously unknown elementary particle and its detection in the particle accelerator would possibly be experimental proof of string theory. Due to mathematical and physical inadequacies, which we will discuss later, this is not yet possible.

There are two types of strings: open and closed strings. Open strings have a start and end point, similar to a piece of thread. Certain modes of vibration can be identified as photons or gluons.

Closed strings are ring-shaped, i.e. closed in a circle. A certain vibration mode can be identified as the graviton, the supposed transmission particle of gravity. The lowest energy vibrational mode is understood as a tachyon, the particle that always flies at above-light speed, moving into the past.

In 1995, in a lecture by Edward Witten at the University of California, the assumption prevailed that the string theories in their present form are only an approximation of the actual theory of everything. To date, five ten-dimensional string theories and one eleven-dimensional supergravity theory have emerged. All these theories should therefore be only approximations of an even higher theory, the M-theory, just as Newton's formulae were an approximation of the theory of relativity. Whether M-theory is the long sought after world-wide formula or just an approximation of an even higher level of scientific knowledge will be demonstrated by the intense research that is currently being done in this field at universities and institutions worldwide. At any rate, M-theory is regarded by physicists as currently the most promising candidate for uniting quantum and relativity theories, thereby taking the theory of everything a step closer.

5. Outlook for the 3rd millennium

The 22nd century

Looking back at the previous chapters we find that we have experienced phenomena that could lead to incredible technologies. Technologies that we have so far banished to the realm of fantasy, science fiction. The idea of travelling across the universe, visiting other times or dimensions sounds tempting, frightening or simply fascinating. Of course, we are still a long way from implementing these technologies. Finally, it took more than 300 years for the first plane to take to the air, even though the theoretical and conceptual foundations go back to Leonardo da Vinci. Da Vinci was a brilliant thinker far ahead of his time. He developed sketches and concepts that may be considered precursors of the helicopter. What he lacked was the engine and mechanical skills needed to bring the complex design to fruition. Today we are in a similar situation. We have an unprecedented wealth of theoretical concepts that could take us into completely new spheres. Today we know that time travel or "jumps" into the space-time structure are quite feasible. At least it allows the general relativity theory to be applied. Likewise, we are in the process of discovering completely new ideas and directions of impact in quantum physics. Perhaps this technological development will lead us into the 3rd millennium, which we have already entered. Maybe it set us on completely different paths.

Research is exploring the structures of nature in increasing depth and is reaching into areas that imply a certain susceptibility as far as we humans are concerned. Let us recall that man is the first living being in the known history of the world capable of annihilating himself. The nuclear missiles stored on our planet were enough to wipe out the majority of the population in one fell swoop. Research and progress also means responsibility. Responsibility to bear, control and, above all, consider the consequences of research. We are now in the second decade of the 3rd millennium. We are slowly beginning to penetrate into the spheres of possibilities of relativity and quantum theories. These are spheres in which even our Earth seems comparatively small. We are also talking about energy-related spheres in which a nuclear bomb blast behaves like a needle stick. The further we venture into the structure of the universe and the more elemental things we explore, the greater the risk we have to accept. Let us recall that the commissioning of the new particle accelerator at CERN already caused a sensation in some areas of the professional world. In Geneva, a particle accelerator had been built for billions of dollars to simulate events similar to those of the Big Bang. Some scientists feared that these experiments might create black holes on Earth. In turn, black holes are phenomena that we still cannot understand, cannot comprehensively explain, and of which we have no practical experience. A certain amount of caution is certainly required in other respects, for example when we begin to experiment in energy spheres that have hitherto been hidden from us in theory. What happens when we randomly create exotic or strange matter in an experiment? Or even a small amount of antimatter? Or a substance or phenomenology that is completely unknown to us so far? Some geologists today argue that exotic matter exists not only in physicists' formulae but could account for some earthquakes that have been inexplicable to date. Caution should be exercised because, in the third millennium, we will be able for the first time to carry out experiments whose outcome we cannot predict. These are experiments in gigantic energy spheres where exotic phenomena suddenly become reality.

As we progress we not only generate new technologies but will also have an increasing obligation to keep repeating the responsibility of meaning and benefit over and over again. From a certain point of view research and science should no longer be seen as an opportunity, but as a danger. It is important to deal distantly and objectively with future technologies and developments. Today, of course, we are still a long way from conducting experiments that push us to the limits of general relativity or quantum physics. Yet it is time to reflect on meaning and nonsense, and not blindly seek progress just because in the past it was the driving impulse of humanity. Rethinking also means having a critical look at the usefulness and meaning of the "Millennium" projects. A modern particle accelerator quickly incurred costs of billions of dollars to build and millions more to use and maintain. A NASA scientist figured the cost of producing an antimatter drive was once around $ 200 billion. The cost of opening even more spheres and visiting other solar systems, galaxies, or the vastness of the universe may be even higher by various factors. Despite all the enthusiasm for the beauty and fascination of the universe and of the world in which we live, we should still remember the proportionality. $ 200 billion. This also means: a lot of money to create a lot of misery and poverty in the world. Perhaps it would be a very worthwhile thought to first solve the problems of the world before embarking on what is probably the greatest exploration of all time. On a journey which takes us into the incredible, unimaginable, fascinating expanses of the universe, space and time. However, it is probably premature to give serious thought to such journeys. History has shown that in very few cases are the technological developments predictable. Just before the outbreak of World War

II, visionaries conjectured that in the 21st century, flying vehicles and ships would populate the cities. In the 1950's, NASA published futuristic conceptual sketches of lunar and space stations that were never built. Later, the terraforming of Mars was added, transforming its atmosphere into a life-friendly space. Even in the 90's, illustrators and authors dreamed of lifts to the moon or manned Mars landings. However, nobody could foresee the really significant technological developments, those that shape the zeitgeist and society. Who would guess that one day the computer and Internet would have had such a big impact? Even industrial insiders estimate the global demand for computers at "about five pieces" almost half a century ago. Nobody guessed how hyperlinks and home pages would change the world, indeed the whole of society, so significantly and enduringly. The really significant developments in the history of science and humanity are mostly unpredictable. So the 22nd century will probably be revolutionary, but probably will not bring forth time machines, interstellar spaceships or teleporters. Yet probably we will see a variety of new technologies that it is impossible to guess at. the past, progress in space was dominated by the American Space Agency NASA and, in the first half of the Cold War, by the Soviet counterpart. Since the first landing on the moon in 1969, government interest in space travel has steadily declined, with a temporary peak in mid-2011 when the planned lunar station or manned voyage to Mars fell victim to US austerity measures. Considering that the public finances of the once glorious space nations are crumbling, it is unlikely that bold space adventures will see a renaissance in the next few years. At least not funded by the state. Silicon Valley companies like SpaceX are trying to privatise space flights. Whether this succeeds will become apparent in the next few years.

5.2 What if ...

... the relativity and quantum theories are wrong?

One morning you are driving to work as the car radio, newspapers and every conceivable channel of communication loudly proclaim the downfall of modern physics:

"Proof: relativity theory and quantum theory wrong!"

At first, you will probably be looking for a free parking space in the company garage, annoyed at the trainee who is pleased to park in your long-term space, you look for another car park and you are annoyed at the colleague who is claiming two spaces with his lease, and last but not least the chewing gum that someone spat on the steps for lack of community spirit. You then quickly get the obligatory coffee and a biscuit - otherwise there is no treat - and finally sit down at your usual workplace and wait for the things that may come your way.

But wait: there was something else. The death knell of modern physics!

You quickly open the Internet and read the news portals up and down. And lo and behold, somebody has actually found proof that modern physics is a modern lie.

What now?

Now. Of course, it is very unlikely that you will be annoyed about the pretty trainee who has dared to go to your regular place. But it is quite possible, not to say likely, that you will one day experience such a headline.

The Strange Universe: Einstein, Quantum Physics and the ToE

For almost a century, we can say with a good conscience that our modern science has been based on the theory of relativity and quantum theory. Two theories that have been repeatedly been proven and confirmed in numerous experiments.

Despite all our efforts to understand and contain the universe in mathematical formulae, we must never forget that what we say about the condition of Nature amounts to pure theories. And these theories have the remarkable characteristic that they are falsifiable, otherwise they are not, according to popular belief, scientific. The mere fact that no contradictions to the theory of relativity or quantum physics have been substantiated so far does not mean that there are no contradictions. For Newtonian mechanics, it took 300 years to realise that this theory has very limited validity. However, it is highly unlikely that such a contradiction will suddenly emerge and turn modern physics upside down. It is should rather be assumed that the two theories will be a limiting case in a superordinate theory. Today we already know that the theory of relativity and quantum physics cannot provide an adequate explanation for extreme phenomena such as black holes. It is therefore to be assumed that these theories are valid only in a certain area and therefore one day will be "falsified".

But do not worry: in this context, falsification means rather that the validity of the two theories is confined to a certain area, not that the theories are completely wrong. If, however, it should turn out that relativity and quantum physics are totally wrong, it would have to be explained why the experimental results so far agree so well with the theoretically expected results. If you come across any scientists on the Internet who claim to have found a contradiction to the two theories, you no longer need to wonder about this. In the last century, apparent contradictions to quantum physics and the theory of relativity surfaced but they have been clarified on closer inspection.

5.3 What came before the big bang?

I have long pondered whether it is appropriate to discuss this question in this book. In this context I speak deliberately of discussion because in principle it is not possible to give a demonstrable answer. On the one hand, this question cannot be scientifically answered seriously because our laws of Nature apply only in the context of Planck's constants. One of these constants is Planck time whereby our laws of Nature become valid only about $5 * 10^{-44}$ seconds after the Big Bang. At smaller time intervals our laws of Nature probably lose their validity. It is therefore fundamentally impossible to describe the moment of birth of our universe on the basis of the natural laws known to us. All attempts in this regard are therefore purely speculative. So a world view of the moment of birth of the universe as a regular observer enjoys in principle the same truth content as a so-called scientific representation. Similarly, it makes no sense to postulate an objective view of what has gone before.

But we can choose a "bottom-up" approach and reduce the question of the origin of the universe to answering the basic question of whether life on earth emerged as a result of random events or was created by chance. We would therefore have arrived at the two most common theories about our origin and the origin of life in general - and hence the fundamental question of whether our world was created by natural coincidences or created by a higher being.

The proponents of the random theory like to put forward a scientific view of things. They assume that life resulting from chemical processes emerged accidentally and without the assistance of a "higher power". In contrast, the theory of creation states that the universe and life were created by a higher power.

5.4 All coincidence?

Basically, we first have to define what "scientific" means: An assumption is scientific if you can observe and verify it. In particular, the assumption must exist independently of the person performing the experiment. Sensational reports from Russia, according to which it is possible to suppress gravity by means of superconductors, are not considered scientific, for example, if the experiment cannot be repeated by other scientists with similar success. Based on this definition of science, some physicists categorically reject world formula theories (such as string theory), as their predictions, at least for the time being, defy any experimental verification (and thus are not falsifiable). Since the assumptions made can neither be observed nor experimentally verified, they argue that these theories are not scientific. This could be countered, however, by saying that Einstein's theory of relativity at the time of publication could be reviewed only to a limited degree. Only the development of more accurate measuring devices and future research facilities would allow the detection of certain aspects such as the postulated gravitational waves.

For a theory on the origin of life or the origin of the universe – and hence all of our known world – to be scientific, it would have to be verifiable, or at least comprehensible. It is clear that any Big Bang is hardly reproducible[58], but a scientific theory would have to describe the moment of origin in accordance with the laws of Nature. We would therefore have to deal with the problem origin and evolution theories have in common: the explanation of the zero moment. Thus although a Big Bang may be inferred as the genesis of the universe, if we reverse the extension of the universe observed with telescopes, the we do not yet have any explanation at all for the Big Bang. Nobody knows how a universe could arise with billions of stars and planets out of nowhere. Nobody knows why the Big Bang happened at all. The continuation of the Big Bang theory, which explains the formation of galaxies, stars, planets and moons, is relatively conclusive. However, the theory lacks the spark that would provide the foundation on which it is based: the beginning. When considering different theories, this is often forgotten or deliberately suppressed. The Big Bang theory lacks a scientific explanation of the moment of birth. It describes very nicely the consequences and from this the development of the universe, but not how it came about. It is as if someone claims to understand a serious illness like cancer, but only knows the symptoms and has no idea what is causing the patient's complaint. Ascribing the origin of the universe to the collision of two universes is as helpful as deriving the genesis of the chicken from the egg. The zero moment problem is merely shifted, but not solved.

The same problem arises with the second essential question of

---

[58] In the tabloids experiments on particle accelerators, such as the CERN in Geneva, tend to be compared with the reproduction of the Big Bang. In fact the energies required for in the experiments cannot of course be even remotely compared with the energies involved in the Big Bang. On the contrary, attempts are being made to allow the smallest particles to collide with a comparatively high level of energy.

origin, namely the emergence of life. The theory of evolution reveals how the different life forms have evolved over millennia and millions of years. Even the theory of evolution, however, lacks a conclusive explanation of how the first living entity originated. Conveniently, you could just ascribe this to coincidence. But even the accidental emergence of a unicellular creature is about as likely as a whole dictionary emerging from an explosion at a printing shop. Living things like dinosaurs or humans are even more complex. A human consists of over 100 trillion cells, that is one hundred thousand billion cells. Every (!) human being consists of more cells than there are grains of sand on the earth or stars in the entire known universe. Even if we allow for random events taking place over billions of years, the chance emergence of complex creatures such as those of a dinosaur or man is about as likely as the emergence of a working Boeing 747 when a tornado races over a junkyard.

Another aspect that contradicts against a chance development is the second law of thermodynamics. This states that entropy can never decrease in a closed system. This means that every natural process always causes more disorder. Without external influences, Nature cannot increase order in a closed system by any process. This in turn means: if we consider a junkyard as a closed system over any length of time, the scrap will never give rise to an operational car without external influences. Instead, the scrap will continue to disintegrate due to chemical processes and the disorder will increase. This thermodynamic principle has not yet been refuted in any experiment and is considered to be similarly secured as energy conservation. Our intuition is also shaped and guided by this principle: if we find a fossilised pot during excavations, we attribute it to the work of an earlier civilisation. The sheer complexity of the shape, order and complexity that makes the clay into a pot make us intuitively aware that this pot was man-made and not random. The same applies to the pyramids in Egypt or the stone circles at Stonehenge. A living being like man is, as you will certainly agree with me, much more complex than a pot of clay. A living being like man is a complex arrangement of trillions of molecules. If we assume that the pot was not created accidentally, how can we assume that man is a random product? And if man is a by-product, how do we know that our excavations - pots, fossils, dinosaur skeletons, or even the pyramids - did not come about by accident (and therefore do not bear witness to past civilisations and living beings)?

The intriguing irony of the Big Bang and Evolution Theory is that both theories claim to put the creation of the universe and of life in a scientific light but require that at least the laws of thermodynamics, which are of a scientific nature, be deprived of their validity. Quite apart from the probability calculations, these theories predominate. These laws form the basis of our understanding of Nature and our research in general. It could even be said that these laws ensure the continuity of our being. If Nature did not follow the second law of thermodynamics, we would only need to wait long enough and accomplishments like a Boeing 747 or the pyramids of Egypt would be purely coincidental. If complex creatures such as animals or humans have come into being by chance, it is quite likely that somewhere in the universe much less complicated things like airplanes or nuclear power plants have come about at random. Or they are the testimony to past civilisations excavated by archaeologists. The Big Bang and evolutionary theory only works if a large portion of accepted science - which is scientific, as reproducibly and experimentally confirmed - is ignored. From this we can only conclude that the Big Bang and evolutionary theory, at least in its present form, cannot be scientific - and thus in turn enjoys no higher priority than any other plausible conjecture or assertion as to how our universe or life on Earth has emerged. The Big Bang and evolutionary theory are therefore in principle no more scientific, for example, than the belief in the creation of the world and life by a higher power

5.5 The Century Puzzles of Physics

Our knowledge of Nature has increased rapidly in recent decades. The development of modern telecommunications and the world of computers has contributed to the acceleration of research and progress, as well as increased international cooperation, in order to tackle problems as significant as the theory of everything with combined forces. However, we face the prospect of further turmoil when we look at the outstanding questions that still exist in the Standard Model of Physics. Our theories are good models with which to explain and calculate many processes and procedures in practice, but we really have not understood quantum mechanics or the theory of relativity. We know a good deal about how they work, but not why. We still have eighteen parameters in the standard model which we have determined through experiments that we can neither explain nor deduce.

The following are some of the most important questions of the Standard Model and Physics. Questions that are discussed and explained in detail in this book. If you succeed in finding a final answer to one of these questions, you will probably go down in the annals of physics and be named in the same sentence as Einstein,

Planck, & Co.

5.5.1 The 18 unknowns

We know that the speed of light in a vacuum is about 300,000 kilometres per second. Why does it assume precisely this value? Is there a connection between the three and our three spatial dimensions? Why does an electron have a mass about 2000 times smaller than a proton, with both having the same amount of electrical charge?

An essential debate about the fundamentals leads us to ask why the fundamental constants of nature assume the precise values that we have determined in experiments. Even if some of the eighteen parameters were configured slightly differently, it would not have been possible for planets and stars to be formed. The universe would be quite different from how we perceive it today. Is there an overarching explanation from which the values of the eighteen parameters can necessarily be derived? Rather like two plus two inevitably equals four on the basis of fundamental mathematical principles?

5.5.2 The Mystery of Gravity

Gravity is fundamentally different from the three other basic forces of nature. It is much weaker, it cannot be shielded, has an infinite range, and all attempts to quantify gravity have failed. In addition, gravity fields slow down time. Some researchers explain the puzzling weakness of gravity by postulating additional dimensions that gravity can penetrate, being the only force of nature capable of this.

But why does gravity differ so much from the other three forces of nature? What secrets are hidden behind the curvature of space and time? Is there antigravity and, if so, how are time paradoxes avoided when travelling back in time? How is gravity transferred from the geometric structure of space-time to masses and energies? Does the graviton, the predicted but not yet proven carrier particle of gravity, exist?

5.5.3 Why a universe of matter?

Since the 1930's we have known that antimatter is the counterpart to matter. In various experiments, researchers have found that nature seems to favour matter over antimatter. However, this so-called CP violation does not explain why our universe is nothing but matter. Even if we peer into the depths of the universe, we have so far found no evidence of the existence of antimatter in free Nature. In fact, it has so far only been possible to produce antimatter in the particle accelerator. But why does our universe consist of matter and not of antimatter? And how is the preference of Nature for matter to be explained?

5.5.4 Where does mass come from?

Over the past century, research has achieved epoch-making advances that increasingly confront us with fundamental questions. What is mass? What is energy? From where does matter gain its mass? Although we concern ourselves with these terms in science, we still do not understand where elementary particles get their mass or how it can be that light-fast particles have no rest mass. Also unexplained is the question of why the elementary particles have

have their precise mass. Perhaps the masses can be attributed to the interaction with the so-called "Higgs" particles. In particle accelerator experiments it is currently being examined whether this "Higgs" particle actually exists, thus enabling a significant gap in the standard model of physics to be closed. If this is successfully demonstrated it would mean that mass is not a fundamental property of the elementary particles but is transferred by the Higgs interaction. This in turn does not answer the question of why certain particles interact with the Higgs field (for example, electrons and protons) and why others do not (for example gravitons, photons). Or to put it another way: why are there massless particles and particles having mass? And what is a massless particle anyway?

5.5.5 How many dimensions are there?

The theory of relativity describes our universe as a four-dimensional space-time with three spatial dimensions and the fourth dimension, time. According to the current state of research there are many indications that our universe actually consists of ten, eleven, or twenty-six dimensions. It is conceivable that every force of nature expresses itself in the form of its own dimension, just as gravity represents a curvature of space-time. The additional space dimensions are wound up or coiled and are therefore invisible to us. In any case, the theory of everything does not seem to be formulated without additional dimensions, with far-reaching consequences for our world view and the future of physics.

5.5.6 Are there other universes?

We do not even understand our Earth, but we still wonder if other universes exist. This question is scientifically justified, and experiments have in fact already been carried out at various universities to prove, for example, the existence of mirror universes. Likewise there are theories, which should be taken seriously, whereby gravity in our space-time is much weaker than all other natural forces because it is weakened by additional spatial dimensions. Different theories, which deal with the inner life of horrid black holes and see the tangled matter disappear into another universe, are fired from similar guns. Today we know that our universe consists of at least four dimensions and possibly several additional, invisible extra dimensions. Given these breathtaking insights, the existence of other universes cannot be categorically excluded.

5.5.7 Does dark energy exist?

Science suggests that an invisible form of energy and mass fills a large part of the universe to explain the observed accelerated expansion of the universe as well as the slowed rotation in outer regions of galaxies. But what produces this dark energy or dark mass is uncertain. Are these neutrinos, the effects of vacuum fluctuations or gravitational forces from other dimensions or universes? Is there even a hitherto undiscovered fifth fundamental force of nature or an error in the formalism of the theory of relativity to which the differences between experiment and theory can be ascribed?

5.5.8 What happens in the black hole?

A black hole is an extreme cosmic phenomenon which, due to its incredible gravity, tears up space-time and makes all matter within its reach vanish behind its event horizon. On the event horizon gravity is so strong that not even light can escape the clutches of attraction. From the formalism of general relativity we also know that time stands still on the event horizon. But what happens behind the event horizon or inside the black hole? Is time flowing backwards, taking the tangled matter into the past? Will the mass be transferred to a new universe or a hyperspace surrounding our four-dimensional space-time?

5.5.9 Where does life come from?

This question is as basic and fundamental as the question of the origin of the universe. When we look at animals, plants, humans - living beings - we look at an arrangement of the smallest elements - cells - in an extreme degree of complexity. No technical device that we can build is anywhere near as complex as the human body or the brain. Can such a complex being as life emerge randomly from simple carbon compounds? Or is there a deeper reason or a higher principle behind life? Could the reason perhaps be that we should populate the earth and deal with such philosophical scientific questions? That we enjoy the complexity and the beauty associated with life?

This question is so fundamental that it extends well beyond the achievements of physics and cannot be answered conclusively, even if one day we decipher the formalism of the theory of everything.

5.5.10 Mysteries of Quantum Physics

In quantum physics, probability waves, eerie remote effects, strange state superpositions and quantum leaps of energy prevail. We have good formalisms to mathematically grasp and describe these amazing phenomena, but we still cannot explain why the outcome of an experiment depends on whether we observe the experiment or not, or how the eerie remote effect can seemingly overcome the speed of light. Nor do we have a conclusive interpretation of the quantum nightmare that tells us the fundamental principle on which this microcosmic chaos rests or how quantum physics and general relativity can be combined to make the universe uniform and free of contradiction.

5.5.11 Is mathematics discovered or invented?

Perhaps you have noticed while reading the chapter on the theory of everything that research in this field depends heavily on mathematical methods. Since the research areas of string theory are experimentally inaccessible, at least current facilities, mathematics is for the time being the only tool that can be used to decipher and understand nature more profoundly. When we look at physics and its laws we quickly realise that they are all mathematical formalisms that describe reality very precisely and comprehensively. Although we are of course used to having natural laws expressed in the form of mathematical equations since our early school days, this very fact is astonishing and leads me to ask whether mathematics is a universal good waiting for its discovery, or whether mathematics is an invention of man which happens to be ideally suited for describing the laws of Nature? The General Theory of Relativity - a physical theory written in the language of mathematics - was discovered by Albert Einstein. If Einstein had not been, it would sooner or later have discovered by another similarly gifted physicist. The theory of relativity describes a behaviour of Nature that exists, whether we can understand it, explain it or even describe it mathematically or not. But what about mathematics? Is this science a useful invention, a tool to express scientific contexts that are constantly evolving and improving? Or is mathematics, so to speak, the natural language in which the laws of Nature are written and which therefore cannot be invented by man but only discovered? Is mathematics perhaps the real law of Nature and physics only an interpretation of it?

Just as fundamentally as the question of where elementary particles gain their masses is the question of whether mathematics as a primitive universal law is waiting to be discovered or must be re-invented by man.

5.5.12 What next?

In science, the quality of a scientist is often measured by the amount of quotes that other researchers share with him. Currently, Edward Witten is one of the most cited scientists in the world. Witten deals with the unification of quantum physics and the theory of relativity, i.e. the discovery of a theory of everything, the so-called string theory. Given the resources and capabilities invested in this area of research worldwide, a breakthrough over the next few years and decades is conceivable. In particular, a breakthrough would involve making a prediction that can be verified by experiment. Because at present the theories of the world are so fundamental that they cannot be falsified by experiments. We might also possibly derive further conclusions from a theory of everything theory as to how we could use quantum and relativity effects in technological developments. But pioneering advances are also conceivable in other areas of physics. Thus the discovery of a high-temperature superconductor, which also exhibits superconducting properties at room temperature, would be a sensation and would permanently change our technology and society. This is because it would make it possible for the first time to store and transmit power effectively and without loss. Modern physics is a sea full of exciting discoveries to be found and understood.

The Strange Universe: Einstein, Quantum Physics and the ToE

Appendix A:
List of Figures

Figure 1 A light clock at 25% of the speed of light
Figure 2 Space-time curvature of the Earth
Figure 3 The shortest route from A to B
Figure 4 A wormhole in space-time
Figure 5 A black hole 600 km away
Figure 6 The photoelectric effect
Figure 7 Double slit experiment - expected result
Figure 8 Double slit experiment - actual result.
Figure 9 The Schrödinger cat
Figure 10 The fundamental building blocks of matter

Copyright notes:
Figure 1   Michael Schmid (GNU license)
Figure 4   AllenMcC (GNU license)
Figure 5   Ute Kraus (GNU license
Figure 9   Dhatfield (GNU licence)

www.ingramcontent.com/pod-product-compliance
Lightning Source LLC
Chambersburg PA
CBHW031613210526
45464CB00004B/1551